Memoirs
of the
American Mathematical Society

Number 904

Rank One Higgs Bundles and Representations of Fundamental Groups of Riemann Surfaces

William M. Goldman
Eugene Z. Xia

May 2008 • Volume 193 • Number 904 (fourth of 5 numbers) • ISSN 0065-9266

American Mathematical Society
Providence, Rhode Island

2000 *Mathematics Subject Classification.* Primary 14H40, 30F30, 57M05, 53C26.

Library of Congress Cataloging-in-Publication Data

Goldman, William Mark.
  Rank one Higgs bundles and representations of fundamental groups of Riemann surfaces / William M. Goldman, Eugene Z. Xia.
    p. cm. — (Memoirs of the American Mathematical Society, ISSN 0065-9266 ; no. 904)
  "May 2008, volume 193, number 904 (fourth of 5 numbers)."
  Includes bibliographical references.
  ISBN 978-0-8218-4136-5 (alk. paper)
  1. Surfaces, Deformation of. 2. Riemann surfaces. 3. Geometry, Differential. 4. Geometry, Algebraic. I. Xia, Eugene Zhu, 1963– II. Title.
  QA648.G65 2008
  516.3′6—dc22   2008060005

Memoirs of the American Mathematical Society

This journal is devoted entirely to research in pure and applied mathematics.

Subscription information. The 2008 subscription begins with volume 191 and consists of six mailings, each containing one or more numbers. Subscription prices for 2008 are US$675 list, US$540 institutional member. A late charge of 10% of the subscription price will be imposed on orders received from nonmembers after January 1 of the subscription year. Subscribers outside the United States and India must pay a postage surcharge of US$38; subscribers in India must pay a postage surcharge of US$43. Expedited delivery to destinations in North America US$53; elsewhere US$130. Each number may be ordered separately; *please specify number* when ordering an individual number. For prices and titles of recently released numbers, see the New Publications sections of the *Notices of the American Mathematical Society*.

Back number information. For back issues see the *AMS Catalog of Publications*.

Subscriptions and orders should be addressed to the American Mathematical Society, P. O. Box 845904, Boston, MA 02284-5904, USA. *All orders must be accompanied by payment.* Other correspondence should be addressed to 201 Charles Street, Providence, RI 02904-2294, USA.

Copying and reprinting. Individual readers of this publication, and nonprofit libraries acting for them, are permitted to make fair use of the material, such as to copy a chapter for use in teaching or research. Permission is granted to quote brief passages from this publication in reviews, provided the customary acknowledgment of the source is given.

Republication, systematic copying, or multiple reproduction of any material in this publication is permitted only under license from the American Mathematical Society. Requests for such permission should be addressed to the Acquisitions Department, American Mathematical Society, 201 Charles Street, Providence, Rhode Island 02904-2294, USA. Requests can also be made by e-mail to reprint-permission@ams.org.

Memoirs of the American Mathematical Society (ISSN 0065-9266) is published bimonthly (each volume consisting usually of more than one number) by the American Mathematical Society at 201 Charles Street, Providence, RI 02904-2294, USA. Periodicals postage paid at Providence, RI. Postmaster: Send address changes to Memoirs, American Mathematical Society, 201 Charles Street, Providence, RI 02904-2294, USA.

© 2008 by the American Mathematical Society. All rights reserved.
Copyright of this publication reverts to the public domain 28 years after publication. Contact the AMS for copyright status.
This publication is indexed in *Science Citation Index*®, *SciSearch*®, *Research Alert*®, *CompuMath Citation Index*®, *Current Contents*®/*Physical, Chemical & Earth Sciences*.
Printed in the United States of America.

∞ The paper used in this book is acid-free and falls within the guidelines established to ensure permanence and durability.
Visit the AMS home page at http://www.ams.org/

10 9 8 7 6 5 4 3 2 1   13 12 11 10 09 08

*To the memory of
Hsieh Po-Hsun*

Contents

Introduction   1

1. Equivalences of deformation theories   4

2. The Betti and de Rham deformation theories and their moduli spaces   5
    2.1. The Betti groupoid   6
    2.2. The de Rham groupoid   8
    2.3. Equivalence of de Rham and Betti groupoids   14

3. The Dolbeault groupoid   17
    3.1. Holomorphic line bundles   17
    3.2. The moduli spaces   21
    3.3. Geometric structure of the Dolbeault moduli space   23

4. Equivalence of de Rham and Dolbeault groupoids   25
    4.1. Construction of the equivalence   25
    4.2. Higgs coordinates   33
    4.3. Involutions   36

5. Hyperkähler geometry on the moduli space   37
    5.1. The quaternionic structure   37
    5.2. The Riemannian metric   39
    5.3. Complex-symplectic structure   40
    5.4. Quaternionization   42

6. The twistor space   43
    6.1. The complex projective line   43
    6.2. The twistor space as a smooth vector bundle   48
    6.3. A holomorphic atlas for the twistor space   49
    6.4. The twistor lines   51
    6.5. The real structure on the twistor space   52
    6.6. Symplectic geometry of the twistor space   53
    6.7. The lattice quotient   55
    6.8. Functions and flows   56

7. The moduli space and the Riemann period matrix   59
   7.1. Coordinates for the Betti moduli space   59
   7.2. Abelian differentials and their periods   60
   7.3. Flat connections   62
   7.4. Higgs fields   64
   7.5. The $\mathbb{C}^*$-action in terms of the period matrix   65
   7.6. The $\mathbb{C}^*$-action and the real points   65

Bibliography   67

Abstract

This expository article details the theory of rank one Higgs bundles over a closed Riemann surface X and their relation to representations of the fundamental group of X. We construct an equivalence between the deformation theories of flat connections and Higgs pairs. This provides an identification of moduli spaces arising in different contexts. The moduli spaces are real Lie groups. From each context arises a complex structure, and the different complex structures define a hyperkähler structure. The twistor space, real forms, and various group actions are computed explicitly in terms of the Jacobian of X. We describe the moduli spaces and their geometry in terms of the Riemann period matrix of X.

This is the simplest case of the theory developed by Hitchin, Simpson and others. We emphasize its formal aspects that generalize to higher rank Higgs bundles over higher dimensional Kähler manifolds.

Received by the editor November 18, 2005.

2000 *Mathematics Subject Classification.* Primary 14H40, 30F30, 57M05, 53C26.

Key words and phrases. Riemann surface, fundamental group, connection, holomorphic structure, Hermitian metric, Jacobi variety, representation variety, hyper-Kähler manifold, Hamiltonian flow.

Goldman gratefully acknowledges partial support by NSF grants DMS-9504764, DMS-9803518, DMS-0103889 and a Semester Research Award from the General Research Board of the University of Maryland in Fall 1998. Xia gratefully acknowledges partial support by National Science Council Taiwan grant NSC 91-2115-M-006-022, 93-2115-M-006-002.

Rank One Higgs Bundles

Introduction

The set of equivalence classes of representations of the fundamental group π of a closed Riemann surface X into a Lie group G is a basic object naturally associated to π and G. Powerful analytic techniques have been employed by Hitchin, Simpson, Corlette and Donaldson et al [**H1, Si1, Si2, Si3**] to understand the global topology and geometry of this object. Rank one Higgs bundles provide a toy model with a more explicit form than general Higgs bundles. This paper expounds this case, emphasizing its formal aspects to isolate and clarify the main ideas and motivate its generalization to higher rank.

We consider three moduli spaces of seemingly different objects: complex characters of the fundamental group π of a closed and orientable surface Σ, flat connections on a trivial line bundle over Σ, and pairs consisting of a holomorphic line bundle over a Riemann surface and a holomorphic 1-form on X. These three moduli problems or *deformation theories* arise in different contexts, but are nonetheless equivalent. Simpson [**Si3**] names these deformation theories Betti, de Rham, and Dolbeault respectively. In rank one, the Betti moduli space is the collection of ordered $2k$-tuples of nonzero complex numbers (k is the genus of Σ). The de Rham moduli space is the quotient

$$\mathrm{H}^1(\Sigma, \mathbb{C})/\mathrm{H}^1(\Sigma, \mathbb{Z}) \cong \mathrm{H}^1(\Sigma, \mathbb{C}/\mathbb{Z}).$$

The Dolbeault moduli space is the cotangent bundle $T^*\mathsf{Jac}(X)$ of the Jacobian of the Riemann surface X.

This paper assumes a basic knowledge of topology (for example, as in Fulton [**Fu**]) and differential, and complex geometry. Our perspective is differential geometric and gauge-theoretic. We assume basic facts about symplectic geometry and Hamiltonian flows (as in Weinstein [**Wei**]). We hope this exposition of the simplest part of deep ideas in algebraic geometry will be useful to topologists and differential geometers interested in moduli spaces of representations of surface groups.

The paper is organized as follows. §1 is a brief introduction of groupoids and their equivalences. §2 constructs the Betti, de Rham, groupoids and moduli spaces and the various structures on these moduli spaces. The main player in the Betti groupoid is a surface group π, whereas the main player in the de Rham groupoid is a closed smooth surface Σ with fundamental group π. §3 develops the Dolbeault groupoid; here Σ is given a conformal structure, making it a Riemann surface X over which we consider various holomorphic objects. The inherent complex structure J on the Betti and de Rham moduli spaces are isomorphic, but J differs from the inherent complex structure I on the Dolbeault moduli space. In §5, these two different structures give rise to a hyperkähler structure on the underlying moduli space. §6 explicitly constructs the *twistor space,* a holomorphic object containing both moduli spaces. All of these constructions apply to the cotangent bundle of an arbitrary principally polarized abelian variety (not necessarily the Jacobian of a curve). In particular they can all be expressed in terms of the *Riemann period matrix* Π of X.

Our story ends (§7) by explicitly describing the moduli spaces in terms of the period matrix. We revisit the Betti moduli space which appears as a product of the real torus

$$\mathsf{Hom}(\pi, \mathrm{U}(1)) \cong T^{2k}$$

with the real symplectic vector space

$$\mathsf{Hom}(\pi, \mathbb{R}^+) \cong \mathrm{H}^1(\Sigma, \mathbb{R}) \cong \mathbb{R}^{2k}.$$

The new structure as the (Dolbeault) moduli space identifies the torus $\mathsf{Hom}(\pi, \mathrm{U}(1))$ as the quotient of $\mathbb{C}^k$ with the lattice $\mathbb{Z}^k + \Pi\mathbb{Z}^k$ corresponding to the columns of Π. The conformal structure of the Riemann surface X determines a complex structure $\mathbb{J}_\Pi$ on $\mathbb{R}^{2k}$ which is easily expressed (7.5.1) in terms of Π. The resulting $\mathbb{C}^*$-action on the moduli space, which determines the full structure of the Dolbeault moduli space and is equivalent to the *Hodge structure* on the moduli space.

Although many features of the rank one case generalize to higher rank, several important technical issues are altogether absent in rank one. We warn the reader of these simplifications. First, since the structure group is abelian, the moduli spaces are naturally groups (or torsors over groups). This strong structure is missing for higher rank Higgs bundles. Endomorphism bundles of line bundles are trivial, but endomorphism bundles are generally nontrivial. In rank one, moduli spaces exist for all objects. One need not remove from consideration "unstable" objects in the sense of Geometric Invariant Theory. Furthermore the main differential equation —- Hitchin's self-duality equation —

decouples, leaving the Higgs field and the holomorphic structure completely independent of each other. This arises from the direct product decomposition $\mathbb{C}^* \cong \mathrm{U}(1) \times \mathbb{R}^+$, which does not generalize to higher rank.

We do not discuss the hyperkähler moment map and quotient construction in this paper, although all the moduli spaces described here can be constructed as complex-symplectic and hyperkähler quotients. The constructions in rank one are particularly simple and familiar. As we do not need this machinery, we do not discuss them here.

The Betti, de Rham and Dolbeaut moduli spaces admit a purely algebraic construction. In the algebraic category, the Betti and de Rham moduli spaces are not isomorphic. However, they are isomorphic as complex analytic Lie groups. Since our approach is complex analytic, we do not discuss this difference in our exposition.

For other points of view and related topics, see [**A, ABCKT, C2, D3, Si1, Si2, Si3, Si4, Si5, T**] and references cited therein.

We thank Nigel Hitchin, Taejung Kim, Weiping Li, Michael Thaddeus and Mike Wolf for their critical reading of this manuscript and numerous helpful suggestions. We are also grateful to Indrianil Biswas, Steve Bradlow, Robert Bryant, Kevin Corlette, Simon Donaldson, Lawrence Ein, Elisha Falbel, Charlie Frohman, Oscar Garcia-Prada, Steve Kudla, François Labourie, John Loftin, John Millson, Niranjan Ramachandran, Jonathan Rosenberg, Carlos Simpson, Domingo Toledo, Richard Wentworth, and Scott Wolpert for helpful conversations.

Notation and terminology. To emphasize the different contexts, we reserve Σ for the smooth surface, and X for the Riemann surface "diffeomorphic to Σ." That is, X is Σ with a conformal structure (which for us is the Hodge $\star$-operator on 1-forms). For constructions involving the differential structure, whether we use Σ or X is a decision on the context. For example, both $\mathcal{A}^*(\Sigma)$ and $\mathcal{A}^*(X)$ are correct notations for the de Rham algebra of smooth differential forms on the surface.

Differential forms and cohomology classes are complex-valued, unless otherwise stated. Tensor products of modules are over $\mathbb{Z}$ unless otherwise stated. If V is a complex vector space, then we denote the real vector space underlying V by $V_\mathbb{R}$. The complex vector space V then consists of the pair $(V_\mathbb{R}, I)$ where $I : V_\mathbb{R} \longrightarrow V_\mathbb{R}$ is the *complex structure*, the $\mathbb{R}$-linear automorphism I satisfying $I^2 = -1$ corresponding to scalar multiplication by $\sqrt{-1}$. The complex vector space $\bar{V}$ *complex-conjugate* to V is $(V_\mathbb{R}, -I)$ where $-I$ is the *opposite* complex structure. Thus an *anti-linear* map $V \longrightarrow W$ into a complex vector space W is a linear map $\bar{V} \longrightarrow W$.

The multiplicative groups of nonzero complex and real numbers are denoted by $\mathbb{C}^*$ and $\mathbb{R}^*$ respectively. The multiplicative group of positive real numbers is denoted $\mathbb{R}^+$. The multiplicative group of unit complex numbers is denoted U(1). We denote the $k \times k$ identity matrix by $\mathbb{I}_k$.

By a *vector bundle* we mean a *smooth complex vector bundle*. If $E \longrightarrow M$ is a vector bundle and N is a submanifold with inclusion map $f : N \hookrightarrow M$, we denote by $E|_N$ the *restriction* of E to N, that is, the vector bundle over N defined as pullback f^*E of E by f.

1. Equivalences of deformation theories

A moduli problem seeks to classify a class of objects up to an equivalence relation, often defined by a group of transformations of the set of objects.

A *deformation theory* (or *transformation groupoid*) (S, G) consists of a category $\mathcal{C}$ defined by a group action as follows. Let $\alpha : G \times S \longrightarrow S$ be a left action of a group G on a set S. The *deformation theory* (S, G) consists of the category $\mathcal{C}$ whose objects form a set $\mathsf{Obj}(\mathcal{C}) = S$ with morphisms

$$x \xrightarrow{g} y.$$

corresponding to triples $(g, x, y) \in G \times S \times S$ such that $\alpha(g, x) = y$.

The identity element $e \in G$ determines, for each object $x \in S$ the identity morphism

$$x \xrightarrow{e} x.$$

The inverse of the morphism

$$x \xrightarrow{g} y$$

is

$$y \xrightarrow{g^{-1}} x$$

and the composition of morphisms

$$x \xrightarrow{g} y \xrightarrow{h} z$$

equals

$$x \xrightarrow{hg} z.$$

In particular every morphism is an isomorphism.

The *moduli set* corresponding to such a deformation theory is the set $\mathsf{Iso}(\mathcal{C})$ of isomorphism classes of objects. The *isotropy group* of an object $x \in \mathsf{Obj}(\mathcal{C})$ is the set $\mathsf{Mor}(x, x)$ consisting of morphisms $x \longrightarrow x$, which has the structure of a group. An *equivalence of categories* is a functor $F : \mathcal{A} \longrightarrow \mathcal{B}$ such that there exists a functor $H : \mathcal{B} \longrightarrow \mathcal{A}$ and natural transformations from the compositions $F \circ H$ and $H \circ F$

to the identity functors of $\mathcal{B}$ and $\mathcal{A}$ respectively. (See Jacobson [**Ja**] or Gelfand-Manin [**GM**], p.28 for discussion of this notion and Goldman-Millson [**GoMi**] for an application closely related to this one.) An equivalence of categories induces a bijection $\mathsf{Iso}(\mathcal{A}) \longrightarrow \mathsf{Iso}(\mathcal{B})$, although in general $\mathsf{Obj}(\mathcal{A})$ and $\mathsf{Obj}(\mathcal{B})$ may be enormously different. For example, each groupoid arising from a group G operating on itself by left-multiplication is equivalent to the groupoid with one object and one morphism.

Equivalent deformation theories yield equivalent moduli sets. However the finer notion of equivalence has further implications— for example isotropy groups of corresponding points in the moduli spaces are isomorphic.

Often the sets $\mathsf{Obj}(\mathcal{A})$ admit additional algebraic or geometric structures, which induce additional structures on $\mathsf{Iso}(\mathcal{A})$. For the examples discussed here, these moduli sets are Lie groups, and the equivalences of deformation theories induces isomorphisms of (real) Lie groups.

Equivalent deformation theories may have different structures. An equivalence of a deformation theory $\mathcal{A}$ with another deformation theory may provide additional structures to $\mathsf{Iso}(\mathcal{A})$. For example, $\mathsf{Hom}(\pi, \mathrm{U}(1))$ inherits the structure of a complex abelian variety from every Riemann surface with fundamental group π.

The following criterion is a useful tool for proving that a functor is an equivalence of categories. A functor $F : \mathcal{A} \longrightarrow \mathcal{B}$ is an *equivalence* if and only if:

- **Surjective on isomorphism classes:** The induced map
  $$F_* : \mathsf{Iso}(\mathcal{A}) \longrightarrow \mathsf{Iso}(\mathcal{B})$$
  is surjective;
- **Full:** For $x, y \in \mathsf{Obj}(\mathcal{A})$, the map
  $$F(x,y) : \mathsf{Mor}(x,y) \longrightarrow \mathsf{Mor}(F(x), F(y))$$
  is surjective;
- **Faithful:** For $x, y \in \mathsf{Obj}(\mathcal{A})$, the map
  $$F(x,y) : \mathsf{Mor}(x,y) \longrightarrow \mathsf{Mor}(F(x), F(y))$$
  is injective.

2. The Betti and de Rham deformation theories and their moduli spaces

This section describes the Betti and de Rham deformation theories. Fix a structure group G. Although much of what is here generalizes to the case that G is a linear algebraic group, we restrict in this paper

to the cases of G being the group $\mathbb{C}^*$ of nonzero complex numbers, the group $U(1)$ of unit complex numbers, or the group $\mathbb{R}^*$ of nonzero real numbers. In what follows, Σ is a compact smooth oriented surface with fundamental group π.

- *The Betti groupoid* whose objects are representations $\pi \longrightarrow G$, with morphisms G;
- *The de Rham groupoid* whose objects are flat connections on a trivial complex line bundle over Σ, with morphisms gauge transformations;

2.1. The Betti groupoid. Denote by $\mathsf{Hom}(\pi, G)$ the set of representations from π to G. The group G acts on representations by conjugation. The *Betti groupoid* is the category having $\mathsf{Hom}(\pi, G)$ as the set of objects and morphisms

$$g : \rho_1 \longrightarrow \rho_2$$

where $g \in G$, $\rho_1, \rho_2 \in \mathsf{Hom}(\pi, G)$ and

$$\rho_2 = \iota_g \circ \rho_1$$

where $\iota_g : G \longrightarrow G$ is the inner automorphism defined by conjugation by g. The Betti groupoid is $(\mathsf{Hom}(\pi, G), G)$.

The surface group π admits a presentation

(2.1.1) $\qquad \langle A_1, B_1, \ldots, A_k, B_k \mid [A_1, B_1] \ldots [A_k, B_k] = 1 \rangle$

where $[A, B]$ denotes $ABA^{-1}B^{-1}$. The map

$$\mathsf{Hom}(\pi, G) \longrightarrow G^{2k}$$
$$\rho \longmapsto (\rho(A_1), \rho(B_1), \ldots, \rho(A_k), \rho(B_k))$$

embeds $\mathsf{Hom}(\pi, G)$ as the Zariski-closed subset of G^{2k} defined by

(2.1.2) $\qquad [\alpha_1, \beta_1] \ldots [\alpha_k, \beta_k] = 1.$

Since G is abelian, it acts trivially on $\mathsf{Hom}(\pi, G)$. Furthermore (2.1.2) is trivially satisfied and

(2.1.3) $\qquad \mathsf{Hom}(\pi, G)/G \cong \mathsf{Hom}(\pi, G) \cong G^{2k}.$

Hence the moduli space $\mathsf{Hom}(\pi, \mathbb{C}^*)$ identifies with $(\mathbb{C}^*)^{2k}$ with a natural complex structure J.

Each $\rho \in \mathsf{Hom}(\pi, G)$ provides $\mathfrak{g}$ with a π-module structure via the adjoint action of G on $\mathfrak{g}$. Since G is abelian, this π-module structure is trivial. The tangent space of $\mathsf{Hom}(\pi, G)$ at ρ identifies with the group cohomology $\mathrm{H}^1(\pi, \mathfrak{g})$. Cup product

$$\mathrm{H}^1(\pi, \mathfrak{g}) \times \mathrm{H}^1(\pi, \mathfrak{g}) \longrightarrow \mathbb{C},$$

defined by

$$\Omega(u,v) := \sum_{i=1}^{g} u(A_i)v(B_i) - u(B_i)v(A_i), \qquad (2.1.4)$$

provides a closed nondegenerate exterior 2-form on $\mathsf{Hom}(\pi, G)$. When $G = \mathbb{C}^*$, this closed exterior 2-form Ω is holomorphic with respect to J. Thus (J, Ω) defines a *complex-symplectic structure* on $\mathsf{Hom}(\pi, \mathbb{C}^*)$.

Each element $C \in \pi$ defines a function

$$\begin{aligned} f_C : \mathsf{Hom}(\pi, G) &\longrightarrow \mathbb{C} \\ \rho &\longmapsto \rho(C). \end{aligned} \qquad (2.1.5)$$

which determines a Hamiltonian flow on $\mathsf{Hom}(\pi, G)$ (Goldman [**G2**]) as follows. Since G is abelian, the function f_C depends only on the homology class of C. By applying an automorphism of π, we may assume that $C = (A_1)^n$ for some $n \geq 0$. The corresponding flow

$$\Phi_t : \mathsf{Hom}(\pi, G) \longrightarrow \mathsf{Hom}(\pi, G)$$

for $t \in \mathbb{R}$ is defined on the generators as follows:

$$\Phi_t(\rho) : \begin{cases} A_1 &\longmapsto \rho(A_1) \\ B_1 &\longmapsto \rho(B_1)e^{int} \\ A_j &\longmapsto \rho(A_j) & \text{if } j > 1 \\ B_j &\longmapsto \rho(B_j) & \text{if } j > 1 \end{cases} \qquad (2.1.6)$$

$\mathsf{Hom}(\pi, \mathbb{C}^*)$ is a complex Lie group under pointwise multiplication of homomorphisms. Namely, if ρ_1, ρ_2 are homomorphisms, then

$$\begin{aligned} \pi &\longrightarrow \mathbb{C}^* \\ \gamma &\longmapsto \rho_1(\gamma)\rho_2(\gamma) \end{aligned} \qquad (2.1.7)$$

is a homomorphism, defining a group structure on $\mathsf{Hom}(\pi, \mathbb{C}^*)$ isomorphic to $(\mathbb{C}^*)^{2k}$.

Two real forms of $\mathbb{C}^*$ are $\mathrm{U}(1)$ and $\mathbb{R}^*$. These subgroups are the fixed point sets of anti-involutions of $\mathbb{C}^*$ defined by

$$\iota_U : z \longmapsto (\bar{z})^{-1}, \qquad \iota_\mathbb{R} : z \longmapsto \bar{z}$$

respectively. These induce real structures on $\mathsf{Hom}(\pi, \mathbb{C}^*)$ whose sets of fixed points are $\mathsf{Hom}(\pi, \mathrm{U}(1))$ and $\mathsf{Hom}(\pi, \mathbb{R}^*)$ respectively. Note that the composition of the anti-involutions is the involution

$$\iota_U \circ \iota_\mathbb{R} : z \longmapsto z^{-1}.$$

2.2. The de Rham groupoid.
The second deformation theory concerns flat connections on a smooth complex line bundle E over a smooth surface Σ. The morphisms are gauge transformations, that is, smooth maps $\Sigma \longrightarrow G$. A flat connection defines a notion of a smooth section being locally constant, or *parallel*. The connection is described as a differential operator which vanishes precisely on locally constant sections.

We restrict to the case when E is the trivial line bundle. As the Betti moduli space, the de Rham moduli space is a complex Lie group of dimension $2k$, where k is the genus of Σ. It furthermore enjoys an invariant complex-symplectic structure as well as two real structures (anti-automorphisms) corresponding to the real forms of $\mathbb{C}^*$.

We first review exterior differential forms, connections, and gauge transformations. For background on fiber bundles and their differential geometry, we refer the reader to Steenrod [**S**] and Kobayashi-Nomizu [**KN**].

Exterior differential calculus. Let $\mathcal{A}^*(\Sigma)$ denote the *de Rham algebra of* Σ, that is, the differential graded algebra of $\mathbb{C}$-valued smooth exterior differential forms on Σ. The operations are *wedge product*

$$\mathcal{A}^k(\Sigma) \times \mathcal{A}^l(\Sigma) \longrightarrow \mathcal{A}^{k+l}(\Sigma)$$

and *exterior derivative*

$$d: \mathcal{A}^k(\Sigma) \longrightarrow \mathcal{A}^{k+1}(\Sigma)$$

satisfying:
- $d(\xi \wedge \eta) = d\xi \wedge \eta + (-1)^k \xi \wedge d\eta$ if $\xi \in \mathcal{A}^k(\Sigma)$;
- $d \circ d = 0$.

Complex-conjugation $\eta \longmapsto \bar{\eta}$ defines an anti-involution of the complex vector space $\mathcal{A}^*(\Sigma)$. Its fixed-point set is the subalgebra $\mathcal{A}^*(\Sigma; \mathbb{R})$ of real differential forms. Its -1-eigenspace consists of the subspace $\mathcal{A}^*(\Sigma; i\mathbb{R})$ of purely imaginary differential forms.

Let E be a (smooth complex) vector bundle over Σ. Let $\mathcal{A}^k(\Sigma; E)$ denote the collection of E-valued exterior k-forms over Σ; then the graded vector space

$$\mathcal{A}^*(\Sigma; E) = \bigoplus_{k \geq 0} \mathcal{A}^k(\Sigma; E)$$

is a graded module over $\mathcal{A}^*(\Sigma)$.

Let E be a trivial line bundle. A *trivialization* τ is a nonvanishing section of E. Then $\mathcal{A}^k(\Sigma; E)$ identifies with $\mathcal{A}^k(\Sigma)$ via

$$\mathcal{A}^k(\Sigma) \longrightarrow \mathcal{A}^k(\Sigma; E)$$
(2.2.1)
$$\eta \longmapsto \eta\tau.$$

Gauge transformations. A *linear gauge transformation* of E is a smooth bundle automorphism $\xi : E \longrightarrow E$ covering the identity map of Σ. Thus for each $x \in S$, the gauge transformation ξ acts on the fiber E_x by scalar multiplication of some scalar $g(x) \in \mathbb{C}^*$. In terms of τ,

$$\xi(\tau) = g \cdot \tau$$

but the smooth map $g : \Sigma \longrightarrow \mathbb{C}^*$ is independent of τ since $\mathbb{C}^*$ is abelian. The group $\mathcal{G}_l(E)$ of linear gauge transformations of E identifies with the space $\mathsf{Map}(\Sigma, \mathbb{C}^*)$ of smooth maps $\Sigma \longrightarrow \mathbb{C}^*$. Similarly, the U(1)-gauge group is the subgroup $\mathsf{Map}(\Sigma, \mathrm{U}(1))$ of smooth U(1)-valued maps.

The fundamental groups of $G = \mathrm{U}(1)$ and $\mathbb{C}^*$ are infinitely cyclic, so a smooth map $g \in \mathsf{Map}(\Sigma, G)$ induces a homomorphism

$$\pi_1(g) : \pi \longrightarrow \mathbb{Z},$$

determining an element of

$$\mathsf{Hom}(\pi, \mathbb{Z}) \cong \mathrm{H}^1(\Sigma, \mathbb{Z}).$$

The resulting group homomorphism $\mathsf{Map}(\Sigma, G) \longrightarrow \mathrm{H}^1(\Sigma, \mathbb{Z})$ induces an isomorphism of the group of connected components

$$\pi_0\big(\mathsf{Map}(\Sigma, G)\big) \cong \mathrm{H}^1(\Sigma, \mathbb{Z}).$$

Its kernel is the identity component $\mathsf{Map}(\Sigma, G)^0$ consisting of null-homotopic maps $\Sigma \longrightarrow G$. Since

$$\mathbb{Z} \hookrightarrow \mathbb{C} \xrightarrow{\mathcal{E}} \mathbb{C}^*$$
$$z \longmapsto \exp(2\pi i z),$$

is a universal covering space, $\mathsf{Map}(\Sigma, \mathbb{C}^*)^0$ identifies with

$$\mathsf{Map}(\Sigma, \mathbb{C}) = \mathcal{A}^0(\Sigma)$$

via the exact sequence

$$\mathsf{Map}(\Sigma, \mathbb{C}) \xrightarrow{\exp_*} \mathsf{Map}(\Sigma, \mathbb{C}^*) \xrightarrow{S} \mathrm{H}^1(\Sigma, \mathbb{Z}).$$

Similarly the identity component of $\mathsf{Map}(\Sigma, \mathrm{U}(1))$ identifies with

$$\mathsf{Map}(\Sigma, i\mathbb{R}) = \mathcal{A}^0(\Sigma, i\mathbb{R}).$$

Connections on vector bundles. A *connection* on E is an operator
$$D : \mathcal{A}^0(\Sigma; E) \longrightarrow \mathcal{A}^1(\Sigma; E),$$
such that
(2.2.2) $$D(fs) = fD(s) + df \wedge D(s).$$
($D(s)$ is called the *covariant differential* of s with respect to D.) Such a map extends to
$$D : \mathcal{A}^p(\Sigma; E) \longrightarrow \mathcal{A}^{p+1}(\Sigma; E)$$
by enforcing the identity
$$D(\alpha \wedge \beta) = D(\alpha) \wedge \beta + (-1)^k \alpha \wedge D(\beta)$$
for any $\alpha \in \mathcal{A}^k(\Sigma)$, $\beta \in \mathcal{A}^l(\Sigma; E)$. We denote the space of all connections on E by $\mathfrak{A}(E)$.

A trivialization τ determines a connection
$$D_0 : \mathcal{A}^0(\Sigma; E) \longrightarrow \mathcal{A}^1(\Sigma; E) :$$
by the rule
(2.2.3) $$D_0(f\tau) := df \wedge \tau.$$
(where $f \in \mathcal{A}^0(\Sigma)$ is a smooth function determining a section $f\tau$ by (2.2.1)). This is the unique connection for which τ is *parallel* (has covariant differential zero).

With respect to this connection, an arbitrary connection D has the form
$$D = D_0 + \eta$$
where $\eta \in \mathcal{A}^1(\Sigma)$ and η acts by exterior multiplication. That is, the covariant differential of a section $s = f\tau \in \mathcal{A}^0(\Sigma; E)$ with respect to D is
$$D(s) = df \wedge \tau + \eta \wedge f\tau \in \mathcal{A}^1(\Sigma; E).$$
In particular the 1-form η is given by
$$D(\tau) = \eta \wedge \tau.$$

The *curvature* of a connection D is the $\mathsf{End}(E)$-valued exterior 2-form $F(D) \in \mathcal{A}^2(\Sigma, \mathsf{End}(E))$ such that for any section s of E,
$$D(D(s)) = F(D) \wedge s.$$
If E is a line bundle, then $\mathsf{End}(E)$ is canonically trivial, so $F(D)$ identifies with an ordinary exterior 2-form.

With respect to the trivialization τ, the curvature is:
$$F(D) = d\eta \in \mathcal{A}^2(\Sigma).$$

Hence a connection D is *flat*, that is, its curvature vanishes, if and only if $d\eta = 0$. Thus the space $\mathcal{F}_l(E)$ of flat connections identifies with the subspace $Z^1(\Sigma) \subset \mathcal{A}^1(\Sigma)$ of closed 1-forms. A line bundle (E, D), where $D \in \mathcal{F}_l(E)$ is a *flat line bundle*.

Gauge transformations and the de Rham groupoid. Let τ be a trivialization and let D_0 be the corresponding connection. If $\xi \in \mathcal{G}_l(E)$ corresponds to a map $g \in \mathsf{Map}(S, \mathbb{C}^*)$, then the action of ξ on a connection $D_0 + \eta$ is given by:

(2.2.4) $$\xi \cdot (D_0 + \eta) := D_0 + \eta + g^{-1}dg$$

where $\eta \in \mathcal{A}^1(\Sigma)$.

To see this, let $\xi \cdot \tau$ be the trivialization obtained by transforming τ by ξ. That is, $\xi \cdot \tau = g\tau$ (scalar multiplication). An arbitrary section of E is given by a scalar multiple $s = f\tau$. Let $D = D_0 + \eta$ be an arbitrary connection. The effect of the transformed connection $\xi \cdot D$ on s is given by:

$$\begin{aligned}(\xi \cdot D)(s) &= D(fg\tau) = df \wedge g\tau + fdg \wedge \tau + fgD(\tau) \\ &= df \wedge (\xi \cdot \tau) + fg^{-1}dg \wedge (g\tau) + fg\eta \wedge \tau \\ &= \left(df + f(g^{-1}dg + \eta)\right) \wedge (\xi \cdot \tau)\end{aligned}$$

(since $\mathbb{C}^*$ is abelian). In particular this action is independent of τ.

Naturality of curvature under gauge transformations

$$F(\xi \cdot D) = \xi^*(F(D))$$

follows, in our context, from the closedness of exact forms and the triviality of the action of $\mathcal{G}_l(E)$ on $\mathcal{A}^2(\Sigma)$:

$$F(\xi \cdot D) = d(\eta + g^{-1}dg) = d\eta = F(D) = \xi^* F(D).$$

Hence the $\mathcal{G}_l(E)$-action preserves curvature. The de Rham groupoid is $(\mathcal{F}_l(E), \mathcal{G}_l(E))$.

The de Rham moduli space. The space $\mathfrak{A}(E)$ of all connections is an affine space modeled on the space $\mathcal{A}^1(\Sigma)$ comprising smooth $\mathbb{C}$-valued 1-forms on Σ. Let D_0 be the connection corresponding to the trivialization τ. Every connection on E is of the form $D_0 + \eta$, where $\eta \in \mathcal{A}^1(\Sigma)$. The objects of the de Rham groupoid — flat connections — comprise an affine subspace of $\mathfrak{A}(E)$.

With τ, the space $\mathcal{F}_l(E)$ of flat connections on E identifies with the vector space $\mathcal{Z}^1(\Sigma)$ of closed 1-forms. We shall always assume a fixed trivialization τ_0 for E, which will provide a basepoint D_0 (an *origin*) for the affine space $\mathfrak{A}(E)$ of connections.

The gauge group $\mathcal{G}_l(E)$ identifies with $\mathsf{Map}(\Sigma, \mathbb{C}^*)$ and its action on $\mathfrak{A}(E)$ is given by (2.2.4).

The $\mathcal{G}_l(E)$-action on $\mathfrak{A}(E)$ decomposes into the action of the identity component $\mathcal{G}_l(E)^0 = \mathsf{Map}(\Sigma, \mathbb{C}^*)^0$ on $\mathfrak{A}(E)$ and the action of $\pi_0(\mathcal{G}_l(E))$ on the quotient. If $g \in \mathcal{G}_l(E)^0$ then $g = \exp f$ for some $f \in \mathcal{A}^0(\Sigma)$, and the action on $\mathfrak{A}(E)$ is given by:
$$D_0 + \eta \xmapsto{g} D_0 + \eta + df.$$
Thus the quotient $\mathcal{F}_l(E)/(\mathcal{G}_l(E)^0)$ is an affine space whose underlying vector space is the cohomology
$$\mathrm{H}^1(\Sigma) := \mathcal{Z}^1(\Sigma)/\mathcal{B}^1(\Sigma).$$
The action of $\pi_0(\mathcal{G}_l(E)) \cong \mathrm{H}^1(\Sigma; \mathbb{Z})$ corresponds to the action of $\mathrm{H}^1(\Sigma, \mathbb{Z})$ on $\mathrm{H}^1(\Sigma)$ via the monomorphism
$$(2.2.5) \qquad i^* : \mathrm{H}^1(\Sigma, \mathbb{Z}) \longrightarrow \mathrm{H}^1(\Sigma)$$
induced by the coefficient inclusion $\mathbb{Z} \xhookrightarrow{i} \mathbb{C}$. In summary:

PROPOSITION 2.2.1. *The de Rham moduli space $\mathcal{F}_l(E)/\mathcal{G}_l(E)$ identifies with the cokernel of the map $\mathrm{H}^1(\Sigma, \mathbb{Z}) \longrightarrow \mathrm{H}^1(\Sigma)$ induced by $\mathbb{Z} \hookrightarrow \mathbb{C}$.*

Therefore the moduli space inherits a complex structure J arising from the operation
$$(2.2.6) \qquad J(\eta) := i\eta$$
on 1-forms. Cup product
$$\mathrm{H}^1(\Sigma) \times \mathrm{H}^1(\Sigma) \longrightarrow \mathbb{C},$$
defined on the level of 1-forms by
$$(2.2.7) \qquad \Omega(\alpha, \beta) := \int_\Sigma \alpha \wedge \beta,$$
induces a nondegenerate exterior 2-form Ω on $\mathcal{F}_l(E)/\mathcal{G}_l(E)$. Since Ω is parallel on this affine space, it is closed. Furthermore, this closed exterior 2-form Ω is holomorphic with respect to J. Thus (J, Ω), defines a complex-symplectic structure on $\mathcal{F}_l(E)/\mathcal{G}_l(E)$.

Every $C \in \pi$ defines a function
$$f_C : \mathcal{F}_l(E) \longrightarrow \mathbb{C}$$
$$\omega \longmapsto \int_C \omega$$
corresponding to the function (2.1.5) on the Betti moduli space, and with a Hamiltonian flow corresponding to (2.1.6).

The identity
$$\Omega(J\alpha, \beta) = \Omega(\alpha, J\beta) = i\Omega(\alpha, \beta)$$

means that Ω is complex-bilinear with respect to J (that is, has type $(2,0)$).

Group structure. The de Rham moduli space $\mathrm{H}^1(\Sigma)/\mathrm{H}^1(\Sigma,\mathbb{Z})$ is a Lie group as a quotient of a vector space by a discrete subgroup. Addition corresponds to tensor product of flat line bundles as follows. The trivialization τ_0 of E determines a trivialization $\tau_0 \otimes \tau_0$ of E which we henceforth identify with τ_0. Let (E, D_1) and (E, D_2) be two flat line bundles with local sections s_1, s_2 respectively. In terms of the trivialization, write $s_i = f_i \tau_0$ where $s_i \in \mathcal{A}^0(\Sigma)$ are functions. The *tensor product flat bundle* $(E, D_1) \otimes (E, D_2)$ is the unique flat bundle $(E, D_1 \otimes D_2)$ for which $s_1 \otimes s_2$ is parallel with respect to $D_1 \otimes D_2$, where s_1 and s_2 are local sections of E parallel with respect to D_1 and D_2 respectively.

Write
$$D_1 = D_0 + \eta_1$$
$$D_2 = D_0 + \eta_2$$
$$D_1 \otimes D_2 = D_0 + \eta$$

where $\eta_1, \eta_2, \eta \in \mathcal{Z}^1(\Sigma)$. Then s_i is D_i-parallel if and only if
$$0 = D_i s_i = df_i + \eta_i f_i$$

for $i = 1, 2$ whence
$$\begin{aligned}(D_1 \otimes D_2)(s_1 \otimes s_2) &= D_1(s_1) \otimes s_2 + s_1 \otimes D_2(s_2) \\ &= \big((df_1 + \eta_1 f_1) f_2 + f_1 (df_2 + \eta_2 f_2)\big)\tau_0 \\ &= D_0(s_1 \otimes s_2) + (\eta_1 + \eta_2) s_1 \otimes s_2.\end{aligned}$$

Thus the 1-form η corresponding to the tensor product $D_1 \otimes D_2$ equals the sum $\eta_1 + \eta_2$.

Real structure. Corresponding to the real structures $\iota_\mathbb{R}$ and ι_U on $\mathbb{C}^*$ are real structures on the de Rham moduli space. Their fixed point sets are *real forms* of the de Rham moduli space, corresponding to subdeformation theories for the real forms $U(1)$ and $\mathbb{R}^*$ of $\mathbb{C}^*$. On the Lie algebra $\mathbb{C}$ of $\mathbb{C}^*$ the anti-involutions $\iota_U, \iota_\mathbb{R}$ respectively induce maps

(2.2.8)
$$\eta \xmapsto{\iota_U} -\bar{\eta}$$
$$\eta \xmapsto{\iota_\mathbb{R}} \bar{\eta}$$

which are related by the composition
$$\iota_U \circ \iota_\mathbb{R} : \eta \longmapsto -\eta.$$

These anti-involutions on the Lie algebras are coefficient maps for anti-involutions on the moduli spaces, denoted by $(\iota_U)_*$ and $(\iota_\mathbb{R})_*$ respectively. The fixed point set of $(\iota_U)_*$ is the moduli space $\mathcal{F}_u(E)/\mathcal{G}_u(E)$ of flat unitary connections. The fixed point set of $(\iota_\mathbb{R})_*$ is the moduli space of flat real connections.

Using the real structure, the complex-symplectic form decomposes into real and imaginary parts

$$(2.2.9) \qquad \Omega = \Omega^{\mathrm{Re}} + i\Omega^{\mathrm{Im}},$$

each of which is a *real symplectic structure* on the de Rham moduli space. They satisfy:

$$\Omega^{\mathrm{Re}}(J\alpha, J\beta) = -\Omega^{\mathrm{Re}}(\alpha, \beta)$$
$$\Omega^{\mathrm{Im}}(J\alpha, J\beta) = -\Omega^{\mathrm{Im}}(\alpha, \beta)$$
$$\Omega^{\mathrm{Im}}(\alpha, \beta) = \Omega^{\mathrm{Re}}(J\alpha, \beta).$$

2.3. Equivalence of de Rham and Betti groupoids.

Let G be $\mathbb{C}^*$. The cases of $\mathbb{R}^*, \mathbb{R}^+$ and $U(1)$ are similar. Let D be a flat connection on the trivial line bundle E over Σ. Let τ_0 denote a trivialization, and write

$$D\tau_0 = \eta\tau_0$$

for a closed 1-form η. Over any smooth path

$$[a,b] \xrightarrow{\gamma} \Sigma$$

parallel transport defines a linear map between the fibers

$$\Pi^{-1}(\gamma(a)) \longrightarrow \Pi^{-1}(\gamma(b))$$

as follows. The induced line bundle γ^*E over $[a,b]$ has an induced flat connection γ^*D as well as an induced trivialization $\tau_0 \circ \gamma$. Thus

$$\gamma^*\eta = g(t)dt$$

for a unique function g on $[a,b]$. Similarly, every section s is

$$s(t) = f(t)(\tau_0 \circ \gamma)$$

for a unique function $f(t)$. The section s is parallel if and only if:

$$0 = (\gamma^*D)s = \bigl(f'(t) + g(t)f(t)\bigr)\, dt \otimes (\tau_0 \circ \gamma),$$

that is,

$$0 = f'(t) + g(t)f(t).$$

This differential equation has unique solution

$$f(t) = \exp\left(-\int_a^t g(s)ds\right)f(a)$$

for an arbitrary initial condition $f(a)$. Starting from
$$s(a) = f(a)\tau_0(\gamma(a)),$$
the corresponding section $s(t) = f(t)\tau_0(\gamma(t))$ is the parallel transport of $s(a)$ along γ. Flatness of D implies that the parallel transport from $\Pi^{-1}\gamma(a)$ to $\Pi^{-1}\gamma(b)$ along γ depends only on the homotopy class of γ relative to its endpoints.

If γ is a *based loop*, that is, $\gamma(a) = \gamma(b) = x_0$, then holonomy defines a homomorphism
$$\tag{2.3.1} \mathsf{hol}_p(D) : \pi_1(\Sigma; x_0) \longrightarrow G.$$
(Compare [**KN**].)

Let $\xi \in \mathcal{G}_l(E)$ be a gauge transformation. Evaluation of ξ at x_0 gives a homomorphism of groups $\mathcal{G}_l(E) \to G$, for which (2.3.1) is equivariant. Together these maps give the *holonomy functor* between the de Rham and the Betti groupoid

THEOREM 2.3.1. *The holonomy functor*
$$\mathsf{hol} : (\mathcal{F}_l(E), \mathcal{G}_l(E)) \longrightarrow (\mathsf{Hom}(\pi, G), G)$$
is an equivalence of deformation theories.

PROOF. Let Σ be a smooth manifold with basepoint $x_0 \in \Sigma$. Let $\pi = \pi_1(\Sigma, x_0)$ be the corresponding fundamental group and $\tilde{\Sigma} \xrightarrow{\Pi} \Sigma$ the corresponding universal covering space. Corresponding to a representation $\rho \in \mathsf{Hom}(\pi, G)$ is a *flat line bundle* $\mathbb{C}_\rho \longrightarrow \Sigma$, defined as follows. The group π acts on the total space $\tilde{\Sigma} \times \mathbb{C}$ of the trivial line bundle over $\tilde{\Sigma}$ by deck transformations on the first factor and via ρ on the second factor:
$$\tag{2.3.2} (\tilde{s}, x) \xmapsto{\gamma} (\gamma \cdot \tilde{s}, \rho(\gamma)x)$$
where $x \in G$. The quotient $(\tilde{\Sigma} \times \mathbb{C})/\pi$ is the total space of a smooth line bundle
$$\mathbb{C}_\rho \xrightarrow{\Pi} \Sigma$$
which carries a natural *flat structure* $\mathfrak{F}_\rho$, a foliation of $\mathbb{C}_\rho$ transverse to the fibration such that the restriction of Π to each leaf of $\mathfrak{F}_\rho$ is a covering space of Σ. The leaves are the projections of the leaves $\tilde{\Sigma} \times \{v\}$ of the product foliation of $\tilde{\Sigma} \times \mathbb{C}$ under the quotient map
$$\tilde{\Sigma} \times \mathbb{C} \longrightarrow \mathbb{C}_\rho.$$
The representation ρ is the *holonomy representation* of $\mathbb{C}_\rho$.

The exterior derivative
$$\tilde{d} : \mathcal{A}^k(\tilde{\Sigma}) \longrightarrow \mathcal{A}^{k+1}(\tilde{\Sigma})$$

induces the trivial connection $\tilde{D}$ on the bundle $(\tilde{\Sigma} \times \mathbb{C})$

$$\tilde{D} : \mathcal{A}^k(\tilde{\Sigma}, \tilde{\Sigma} \times \mathbb{C}) \longrightarrow \mathcal{A}^{k+1}(\tilde{\Sigma}, \tilde{\Sigma} \times \mathbb{C})$$

by considering sections as $\mathbb{C}$-valued functions and then taking exterior derivative. The π-action via ρ is equivariant with respect to $\tilde{D}$. Since flatness is a local condition, $\tilde{D}$ descends to a flat connection D on $(\tilde{\Sigma} \times \mathbb{C})/\pi$

A local section s is *parallel* if and only if $D(s) = 0$ if and only if its image lies in a leaf of $\mathfrak{F}_\rho$. Equivalently, a lift of s to $\tilde{\Sigma}$ has constant projection under the Cartesian projection

$$\tilde{\Sigma} \times \mathbb{C} \longrightarrow \mathbb{C}$$

defining the foliation.

The *covering homotopy theorem* for fiber bundles implies that any continuous path $\rho_t \in \mathsf{Hom}(\pi, G)$, where $0 \leq t \leq 1$, induces an isomorphism of line bundles

$$\mathbb{C}_{\rho_0} \longrightarrow \mathbb{C}_{\rho_1}.$$

Since $\mathsf{Hom}(\pi, G)$ is an $\mathbb{R}$-algebraic set, its connected components are path-connected and the topological type of the bundle $\mathbb{C}_\rho$ depends only on the connected component of $\mathsf{Hom}(\pi, G)$ containing ρ. If $\mathsf{Hom}(\pi, G)$ is connected, then any representation can be connected to the trivial representation by a continuous path, and therefore defines the trivial bundle. The existence of such a D for each $\rho \in \mathsf{Hom}(\pi, G)$ shows that hol_p is surjective on isomorphism classes.

Next we show hol_p is faithful and full. Recall that (2.3.1) is equivariant with respect to homomorphism $\mathcal{G}_l(E) \to G$. Hence $\mathcal{G}_l(E)$-equivalent connections give rise to G-equivalent representations. Suppose

$$D_1, D_2 \in \mathcal{F}_l(E).$$

There are two cases, depending on whether or not D_1, D_2 are $\mathcal{G}_l(E)$-equivalent. If D_1, D_2 are not $\mathcal{G}_l(E)$-equivalent, then $\mathsf{Mor}(D_1, D_2)$ is empty. Otherwise, there exists $\xi \in \mathcal{G}_l(E)$ corresponding to a map $g \in \mathsf{Map}(X, G)$ such that

$$D_2 = D_1 + g^{-1}dg$$

and if ξ_1 is another gauge transformation corresponding to a map $g_1 \in \mathsf{Map}(X, G)$ such that

$$D_2 = D_1 + g_1^{-1}dg_1,$$

then $g_1 = gc$, where c is a constant map. We denote the subgroup of $\mathcal{G}_l(E)$ corresponding to constant maps $X \longrightarrow G$ by G. Hence $\mathsf{Mor}(D_1, D_2)$ corresponds to the coset $g \cdot G$.

Now suppose that $D_1, D_2 \in \mathcal{F}_l(E)$ and $\mathsf{hol}_p(D_i) = \rho_i$ for $i = 1, 2$. If D_1 is not $\mathcal{G}_l(E)$-equivalent to D_2, then then $\mathsf{Mor}(\rho_1, \rho_2) = \emptyset$. Otherwise, $\rho_1 = g\rho_1 g^{-1} = \rho_2$ where $g \in G$. $\mathsf{Mor}(\rho_1, \rho_2) = gGg^{-1} = G$. Then we have an isomorphism

$$g^{-1} : \mathsf{Mor}(D_1, D_2) \longrightarrow \mathsf{Mor}(\rho_1, \rho_2).$$

In both cases, hol_p induces an isomorphism

$$\mathsf{Mor}(D_1, D_2) \longrightarrow \mathsf{Mor}(\mathsf{hol}_p(D_1), \mathsf{hol}_p(D_2))$$

as desired. □

For a careful treatment of holonomy for higher ranks and in general, see Kobayashi-Nomizu [**KN**] and Goldman-Millson [**GoMi**]. Theorem 2.3.1 is stated and proved in [**GoMi**].

The holonomy representation of a tensor product of flat line bundles is the product of the holonomy representations, as defined in (2.1.7). Therefore the corresponding isomorphism of moduli spaces is an isomorphism of groups.

If $D = D_0 + \eta$ is a flat connection, then its holonomy is the homomorphism

$$\pi \longrightarrow G$$
$$\gamma \longmapsto \exp \int_\gamma \eta$$

which depends holomorphically on D. Thus the isomorphism of moduli spaces is an isomorphism of complex Lie groups. This isomorphism is also symplectomorphic with regard to the complex-symplectic structures Ω defined above.

3. The Dolbeault groupoid

Let X be a Riemann surface diffeomorphic to Σ. With this extra structure is associated a third deformation theory, the *Dolbeault groupoid*. In this section we define this deformation theory and relate its moduli space to the Jacobian $\mathsf{Jac}(X)$.

3.1. Holomorphic line bundles. We begin by reviewing differential calculus on a Riemann surface. A holomorphic structure is defined as an extension D'' of the $\overline{\partial}$-operator to sections of a line bundle. For line bundles, a Higgs field is a holomorphic 1-form on X. A *Higgs pair* is a holomorphic structure together with a Higgs field.

Exterior calculus on Riemann surface. The Hodge $\star$-operator on 1-forms
$$\star : \mathcal{A}^1(X) \longrightarrow \mathcal{A}^1(X)$$
describes the conformal structure on X as follows. Let $\mathbb{J} : TX \longrightarrow TX$ denote the complex structure on TX: for each $x \in X$, multiplication by $\sqrt{-1}$ is an $\mathbb{R}$-linear automorphism $\mathbb{J}_x$ of the (real) tangent space $(T_xX)_{\mathbb{R}}$ with
$$\mathbb{J}_x \circ \mathbb{J}_x = -1.$$
A 1-form $\eta \in \mathcal{A}^1(X)$ defines an $\mathbb{R}$-linear map $T_xX \longrightarrow \mathbb{C}$. The Hodge $\star$-operator is the map on $\mathcal{A}^1(X)$ induced by $\mathbb{J}$, that is, define $\star\eta \in \mathcal{A}^1(X)$ as the composition
$$T_xX \xrightarrow{\mathbb{J}^{-1}} T_xX \xrightarrow{\eta} \mathbb{C}.$$
(When X is given a conformal Riemannian metric, this operator agrees with the Riemannian $\star$-operator. For 1-forms on a 2-manifold, $\star$ is independent of the Riemannian metric.)

Thus $\star$ defines an endomorphism of the complex vector space $\mathcal{A}^1(X)$ with $\star\star = -1$, so its eigenvalues are $\pm i$. Furthermore $\mathcal{A}^1(X)$ decomposes as a direct sum of its i-eigenspace and $(-i)$-eigenspace respectively:
$$\mathcal{A}^1(X) = \mathcal{A}^{1,0}(X) \oplus \mathcal{A}^{0,1}(X)$$
and the summands are complex-conjugates of each other.

Composing the exterior derivative $d : \mathcal{A}^0(X) \longrightarrow \mathcal{A}^1(X)$ with projections of $\mathcal{A}^1(X)$ onto $\mathcal{A}^{1,0}(X)$ and $\mathcal{A}^{0,1}(X)$ defines operators
$$\partial : \mathcal{A}^0(X) \longrightarrow \mathcal{A}^{1,0}(X)$$
$$\overline{\partial} : \mathcal{A}^0(X) \longrightarrow \mathcal{A}^{0,1}(X)$$
with $d = \partial + \overline{\partial}$.

A *holomorphic 1-form* is a $(1,0)$-form which is closed; a closed $(0,1)$-form is an *anti-holomorphic 1-form*. A *harmonic 1-form* is the sum of a holomorphic 1-form and an antiholomorphic 1-form. The spaces of holomorphic 1-forms, antiholomorphic 1-forms, and harmonic 1-forms, denoted by $\mathcal{H}^{1,0}(X)$, $\mathcal{H}^{0,1}(X)$, and $\mathcal{H}^1(X)$ respectively, satisfy:
$$\mathcal{H}^1(X) = \mathcal{H}^{0,1}(X) \oplus \mathcal{H}^{1,0}(X).$$
where the two summands are complex-conjugates of each other.

Later we need the following basic facts, which can be found in, for example, Farkas-Kra [**FK**], Forster [**Fo**] (Theorem 19.9, p.156, Theorem 19.12, p.159) Griffiths-Harris [**Gr**], Jost [**Jo**],§§3.3,5.2, Wells [**Wel**].

- Let $\eta \in \mathcal{A}^{0,1}(X)$. Then there exists a unique antiholomorphic 1-form $\eta_0 \in \mathcal{H}^{0,1}(X)$ and a function $f \in \mathcal{A}^0(X)$ such that
$$\eta = \eta_0 + \overline{\partial} f$$

- Let $\eta \in \mathcal{Z}^1(X)$ be a closed 1-form. Then there exists a unique harmonic 1-form η_0 and a function $f \in \mathcal{A}^0(X)$ such that
$$\eta = \eta_0 + df.$$

These standard facts form the analytic foundation for the theory expounded here.

Holomorphic structures. Just as a connection on a vector bundle provides a differential criterion characterizing sections to be locally constant, or *parallel*, a holomorphic structure on a vector bundle over a complex manifold provides a notion of a section to be holomorphic.

From a preferred class of *holomorphic local sections,* one can find an atlas of local trivializations of the vector bundle such that the transition functions on overlapping coordinate patches are holomorphic. We do not take that approach here, instead referring to Gunning[**Gu1**], Kobayashi [**K**].

The trivialization τ_0 determines a holomorphic structure, such that $s = f\tau_0$ is a holomorphic section defined on an open set U if and only if f is a holomorphic function on U. In particular holomorphic local sections solve the *Cauchy-Riemann equation*.

Holomorphic structures are conveniently described by differential operators of degree 1
$$D'' : \mathcal{A}^{p,q}(X;E) \longrightarrow \mathcal{A}^{p,q+1}(X;E)$$
which satisfy
$$D''(f \cdot s) = \overline{\partial} f \wedge s + f \cdot D''(s)$$
for $f \in \mathcal{A}^0(X)$. Locally every holomorphic structure admits holomorphic sections, and are thus equivalent to the standard one D_0'' (which arises from $\overline{\partial}$ and the trivialization). This follows from the solvability of the inhomogeneous Cauchy-Riemann equation. See Atiyah-Bott [**AB**], §5 (pp. 554–55) or Kobayashi [**K**], Chapter 1, Propositions 3.5–3.6, p.9 for further discussion.

Denote by D_0' and D_0'' the (1,0)- and (0,1)- parts of D_0 respectively:

$$D_0' = (D_0)^{1,0}$$
$$D_0'' = (D_0)^{0,1}$$

so that
$$D_0 = D_0' + D_0''.$$

With respect to the trivialization τ_0, a *holomorphic structure* on E is an operator
$$D'' = D_0'' + \Psi$$
(where $\Psi \in \mathcal{A}^{0,1}(X)$), which operates on smooth sections by:
$$\mathcal{A}^{0,0}(X, E) \xrightarrow{D''} \mathcal{A}^{0,1}(X, E)$$
$$f\tau_0 \longmapsto (\overline{\partial}f + f\Psi)\tau_0.$$

(When $\dim(X) > 1$, holomorphic structures must satisfy an extra integrability condition. Compare Kobayashi [**K**].)

Denote the space of all holomorphic structures on E by $\mathsf{Hol}(E)$. The linear gauge group $\mathcal{G}_l(E)$ acts on $\mathsf{Hol}(E)$ by

(3.1.1) $$D_0'' + \Psi \xmapsto{\xi} D_0'' + \Psi + g^{-1}\overline{\partial}g.$$

Denote the resulting deformation theory by $(\mathsf{Hol}(E), \mathcal{G}_l(E))$.

Higgs fields. If (E, D'') is a holomorphic vector bundle over X, then a *Higgs field* on (E, D'') is a $(1,0)$-form Φ on X taking values in the endomorphism bundle $\mathsf{End}(E)$. The Higgs field Φ is required to be *holomorphic* with respect to the holomorphic structure on $T^*M \otimes \mathsf{End}(E)$ induced by the complex structure on X and the holomorphic structure D'' on E. Thus a Higgs field is a holomorphic bundle map $TX \longrightarrow \mathsf{End}(E)$. (In higher dimensions, the Higgs field is required to satisfy the integrability condition $\Phi \wedge \Phi = 0$, which automatically holds when X is 1-dimensional.)

A *Higgs pair* (or *Higgs bundle*) is a pair $((E, D''), \Phi)$ where (E, D'') is a holomorphic vector bundle and Φ is a Higgs field on (E, D'').

When E is a holomorphic line bundle, $\mathsf{End}(E)$ is trivial and a Higgs field is a holomorphic 1-form. Thus a rank one Higgs pair is just a holomorphic structure D'' together with a holomorphic 1-form. Thus the space of Higgs pairs (E, D'', Φ) is

(3.1.2) $$\mathsf{Higgs}(E) = \mathsf{Hol}(E) \times \mathcal{H}^{1,0}(X).$$

The linear gauge group $\mathcal{G}_l(E)$ acts on $\mathsf{Hol}(E)$ by (3.1.1) and on Higgs fields by conjugation. Since multiplication in $\mathsf{End}(E)$ is commutative, $\mathcal{G}_l(E)$ acts trivially on $\mathcal{H}^{1,0}(X)$. Denote the resulting groupoid by $(\mathsf{Higgs}(E), \mathcal{G}_l(E))$.

The groupoid $(\mathsf{Higgs}(E), \mathcal{G}_l(E))$ contains the subgroupoid of holomorphic line bundles $(\mathsf{Hol}(E), \mathcal{G}_l(E))$ as Higgs pairs whose Higgs fields are identically zero.

3.2. The moduli spaces.

Now we describe the Dolbeault moduli spaces of rank one holomorphic bundles and Higgs bundles over a compact Riemann surface X. These moduli spaces will identify with spaces of unitary and linear representations of $\pi_1(X)$ respectively.

Moduli of holomorphic structures. The space $\mathsf{Hol}(E)$ of all holomorphic structures is an affine space modeled on $\mathcal{A}^{0,1}(X)$. If D_0'' is the holomorphic structure corresponding to a trivialization, then every holomorphic structure is of the form $D_0'' + \Psi$, where $\Psi \in \mathcal{A}^{0,1}(X)$. As for connections on a trivial line bundle E, we fix a trivialization and its corresponding holomorphic structure. The affine spaces have natural vector space structures, in which vector addition corresponds to tensor product of holomorphic line bundles.

The $\mathcal{G}_l(E)$-action on $\mathsf{Hol}(E)$ is given by

$$(3.2.1) \qquad \Psi \mapsto \Psi + g^{-1}\bar{\partial}g,$$

where $\Psi \in \mathcal{A}^{0,1}(X)$ and $g \in \mathsf{Map}(X, \mathbb{C}^*)$.

The $\mathcal{G}_l(E)$-action on $\mathsf{Hol}(E)$ decomposes into the action of the identity component

$$\mathcal{G}_l(E)^0 = \mathsf{Map}(X, \mathbb{C}^*)^0$$

on $\mathsf{Hol}(E)$ and the action of $\pi_0(\mathcal{G}_l(E))$ on $\mathsf{Hol}(E)/(\mathcal{G}_l(E)^0)$. If $g \in \mathcal{G}_l(E)^0$ then $g = \exp f$ for some $f \in \mathcal{A}^0(X)$, and the action on $\mathsf{Hol}(E)$ is given by:

$$D_0'' + \Psi \xmapsto{g} D_0'' + \Psi + \bar{\partial}f.$$

Thus the quotient $\mathsf{Hol}(E)/(\mathcal{G}_l(E)^0)$ is an affine space whose underlying vector space is the quotient $\mathcal{A}^{0,1}(X)/\bar{\partial}\mathcal{A}^0(X)$, the *Dolbeault cohomology group*. The Hodge decomposition into antiholomorphic forms and $\bar{\partial}$-exact forms (§3)

$$\mathcal{A}^{0,1}(X) = \mathcal{H}^{0,1}(X) \oplus \bar{\partial}\mathcal{A}^0(X)$$

gives an isomorphism of $\mathcal{A}^{0,1}(X)/\bar{\partial}\mathcal{A}^0(X)$ with the space $\mathcal{H}^{0,1}(X)$ of antiholomorphic 1-forms on X, which we henceforth denote as V. Thus $\mathsf{Hol}(E)/(\mathcal{G}_l(E)^0)$ identifies with the complex vector space $V = \mathcal{H}^{0,1}(X)$. The dimension of V equals the genus k of X.

The action of $\pi_0(\mathcal{G}_l(E)) \cong \mathrm{H}^1(X;\mathbb{Z})$ corresponds to the action of $\mathrm{H}^1(X,\mathbb{Z})$ on $\mathcal{H}^{0,1}(X)$ by translation via the composition map

$$(3.2.2) \qquad \mathrm{H}^1(X,\mathbb{Z}) \xrightarrow{i^*} \mathrm{H}^1(X) \xrightarrow{p_1} \mathcal{H}^{0,1}(X) = V$$

where i^* is as in (2.2.5) and p_1 is the projection to antiholomorphic 1-forms. The image of $\mathrm{H}^1(X,\mathbb{Z})$ is a lattice $L \subset V$ of rank

$$2k = \dim(V_\mathbb{R}).$$

The quotient $\mathsf{Jac}(X) := V/L$ is a complex torus, the *Jacobian* of X.

The *Dolbeault isomorphism* identifies $\mathrm{H}^1(X,\mathcal{O})$ with $\mathcal{H}^{0,1}(X)$, where $\mathcal{O}$ is the sheaf of germs of holomorphic functions on X. Under the Dolbeault isomorphism, the above homomorphism
$$\mathrm{H}^1(X,\mathbb{Z}) \longrightarrow \mathcal{H}^{0,1}(X)$$
identifies with the map induced by the inclusion $\mathbb{Z} \longrightarrow \mathcal{O}$ of the sheaf of germs of locally constant $\mathbb{Z}$-valued functions. Its cokernel then identifies with the kernel of the homomorphism $\mathrm{H}^1(X,\mathcal{O}^*) \longrightarrow \mathrm{H}^2(X,\mathbb{Z})$ induced by the sheaf homomorphism
$$\mathcal{O} \xrightarrow{\varepsilon} \mathcal{O}^*$$
$$f \longmapsto \exp(2\pi i f),$$
by the exactness of
$$0 \longrightarrow \mathbb{Z} \longrightarrow \mathcal{O} \xrightarrow{\varepsilon} \mathcal{O}^*$$
where $\mathcal{O}^*$ is the multiplicative sheaf of germs of nonvanishing holomorphic functions.

Just as the tensor product of flat line bundles corresponds to addition of closed 1-forms, the tensor product of holomorphic line bundles corresponds to addition of antiholomorphic 1-forms, that is, to addition in V. Namely if $D_i'' = D_0'' + \Psi_i$, then the tensor product holomorphic structure is
$$D_1'' \otimes D_2'' = D_0'' + \Psi_1 + \Psi_2.$$
This multiplicative structure agrees with the multiplicative structure induced by the multiplicative structure on $\mathcal{O}^*$.

Moduli of Higgs pairs. Let $(D'', \Phi) \in \mathsf{Higgs}(E)$ be a Higgs pair. Its equivalence class is recorded by the equivalence class of the holomorphic bundle (E, D'') in $\mathsf{Jac}(X)$ and the Higgs field Φ in $\mathcal{H}^{1,0}(X)$.

PROPOSITION 3.2.1. *The moduli space* $\mathsf{Higgs}(E)/\mathcal{G}_l(E)$ *of Higgs pairs identifies with the product*
$$\mathsf{Jac}(X) \times \mathcal{H}^{1,0}(X).$$

Complex-conjugation maps the space $\mathcal{H}^{1,0}(X)$ of Higgs fields to
$$\mathcal{H}^{0,1}(X) = \bar{V}.$$
Thus we identify the space of Higgs fields with $\bar{V}$ and Proposition 3.2.1 implies that $\mathsf{Higgs}(E)/\mathcal{G}_l(E)$ is the cokernel of

(3.2.3) $$L \longrightarrow V \times \bar{V},$$

that is, $V/L \times \bar{V}$.

We denote by I the complex structure on $\mathsf{Higgs}(E)/\mathcal{G}_l(E)$ arising from the complex structures on V and $\bar{V}$. In terms of the parameters $\Psi \in V$ and $\Phi \in \bar{V}$ this complex structure is:

$$\Psi \xrightarrow{I} i\Psi$$

(3.2.4)
$$\Phi \xrightarrow{I} i\Phi.$$

Similarly $T^*\mathsf{Jac}(X)$ is an abelian group via a tensor product construction on Higgs pairs. Namely define the tensor product $(D_1'', \Phi_1) \otimes (D_2'', \Phi_2)$ to have holomorphic structure $D_1'' \otimes D_2''$ and Higgs field $\Phi_1 + \Phi_2$.

3.3. Geometric structure of the Dolbeault moduli space.

The conformal structure of X induces a rich geometry to the Dolbeault moduli space $\mathsf{Higgs}(E)/\mathcal{G}_l(E)$. This space is a complex manifold under the complex structure I induced from X, which is the total space of a vector bundle over $\mathsf{Jac}(X)$. This vector bundle is naturally isomorphic to the *cotangent bundle* $T^*\mathsf{Jac}(X)$, which has a natural complex-symplectic structure. Furthermore the Riemannian metric of $\mathsf{Jac}(X)$ defines a proper exhaustion function vanishing on the zero-section $\mathsf{Jac}(X)$, and whose gradient flow defines a deformation-retraction to the zero-section. In rank two, Hitchin [**H1**] used a similar technique to compute the Betti numbers of the Higgs bundle moduli space.

Cotangent bundles. We identify $\mathsf{Higgs}(E)/\mathcal{G}_l(E)$ with the *cotangent bundle* $T^*\mathsf{Jac}(X)$ using a Riemannian metric on $\mathsf{Higgs}(E)/\mathcal{G}_l(E)$ associated to X as follows.

The natural Hermitian form on $\mathcal{A}^1(X)$ defined by

(3.3.1)
$$\langle \alpha, \beta \rangle := \int_X \alpha \wedge \star \bar{\beta}$$

is nondegenerate, positive definite on $\mathcal{A}^{0,1}(X)$ and negative definite on $\mathcal{A}^{1,0}(X)$. Its restriction to $V = \mathcal{H}^{0,1}(X)$ defines an isomorphism $\bar{V} \longrightarrow V^*$ of complex vector spaces. The tangent space of $\mathsf{Jac}(X) = V/L$ at any point identifies with V. By the isomorphism defined by $\langle , \rangle$, the cotangent space of $\mathsf{Jac}(X)$ identifies with $\bar{V}$. Thus

$$V/L \times \bar{V} \cong V/L \times V^* \cong T^*\mathsf{Jac}(X).$$

The real part of $\langle , \rangle$ defines Riemannian metrics on V and $V^* \cong \bar{V}$ which in turn defines a Riemannian metric g on $T^*\mathsf{Jac}(X)$.

The cotangent bundle T^*M of any complex manifold M enjoys a natural complex-symplectic structure Ω_{T^*}, In fact, Ω_{T^*} is *exact*, that is, there exists a holomorphic 1-form α_{T^*} on M such that $\Omega_{T^*} = -d\alpha_{T^*}$. Let $\Pi : T^*M \longrightarrow M$ denote projection and $d\Pi : T(T^*M) \longrightarrow TM$

denote its differential. Let $u \in T^*M$ be a covector at $\Pi(u) \in M$. The value of α_{T^*} at u is defined as:
$$(\alpha_{T^*})_{(u)} : v \longmapsto u(d\Pi(v))$$
for a tangent vector $v \in T_u(T^*M)$. Then
$$\Omega_{T^*} = -d\alpha_{T^*}$$
is a complex-symplectic structure on T^*M. This construction is natural, so any biholomorphism of M preserves α_{T^*} and Ω_{T^*}.

For a complex torus, the complex-symplectic-structure Ω_{T^*} is parallel. If $\mathsf{q} = (q_1, \ldots, q_n)$ are local coordinates on M, with corresponding coordinates $\mathsf{p} = (p_1, \ldots, p_n)$ on the fibers, then
$$\alpha_{T^*} = \sum_{j=1}^n p_i dq_i$$
(3.3.2)
$$\Omega_{T^*} = \sum_{j=1}^n dq_i \wedge dp_i.$$

The space $T^*\mathsf{Jac}(X)$ supports a holomorphic completely integrable Hamiltonian system. The *Hitchin map*

(3.3.3)
$$\begin{array}{c}\mathsf{Higgs}(E)/\mathcal{G}_l(E) \xrightarrow{\mathbf{H}} \mathcal{H}^{1,0}(X) \\ [(D'', \Phi)] \longmapsto \Phi \end{array}$$

associates to a pair (D'', Φ) its Higgs field. In terms of (3.3.2), this is the projection
$$(\mathsf{q}, \mathsf{p}) \longmapsto \mathsf{p}.$$
The complex Hamiltonian vector field with potential p_j is the parallel vector field $\frac{\partial}{\partial q_j}$ which generates translation in the q_j-coordinate. The Hitchin map is a holomorphic moment map for a holomorphic Hamiltonian $\mathbb{C}^k$-action, whose orbits are sections of $T^*\mathsf{Jac}(X)$ corresponding to the trivialization $\mathbf{H}$, where k is the genus of X. Equivalently these parallel copies of $\mathsf{Jac}(X)$ are the preimages of $\mathbf{H}$. Although in this case this construction is elementary, its generalization to higher rank is highly nontrivial and revealing. See Hitchin [**H4**].

On $T^*\mathsf{Jac}(X) = V/L \times \bar{V}$, this structure arises from the linear complex-symplectic structure on the vector space
$$V \oplus \bar{V} = \mathcal{H}^{0,1}(X) \oplus \mathcal{H}^{1,0}(X)$$
defined by

(3.3.4)
$$\Omega_{T^*}\big((\Psi_1, \Phi_1), (\Psi_2, \Phi_2)\big) := \langle \Psi_1, \overline{\Phi_2} \rangle - \langle \Psi_2, \overline{\Phi_1} \rangle.$$

Functions and flows on the Dolbeault moduli space. The *energy function*

$$T^*\mathsf{Jac}(X) \xrightarrow{e} \mathbb{R}$$

(3.3.5) $$(\mathsf{q}, \mathsf{p}) \longmapsto \frac{1}{2}\langle \mathsf{p}, \mathsf{p}\rangle$$

vanishes precisely on the zero-section $\mathsf{Jac}(X)$, which contains the only critical points of e. These critical points are minima. Let ∇e denote the gradient vector field of e with respect to the Riemannian metric g. Then the flow of $-\nabla(e)$ defines a deformation retraction of $T^*\mathsf{Jac}(X)$ to its zero-section $\mathsf{Jac}(X)$, from which one can determine the homotopy type of the moduli space. Hitchin [**H1**] uses a similar technique to determine the topology of some moduli spaces in rank two. However, in higher rank, not all critical points are minima, and the calculation is considerably more complicated. For further applications see [**BGG1, BGG2, Go1, Go2, Go2, X0, X1, X2**].

The vector field $I\nabla(e)$ is the Hamiltonian vector field having potential e with respect to the Kähler form on $T^*\mathsf{Jac}(X)$. It generates the Hamiltonian circle action:

(3.3.6) $$S^1 \times T^*\mathsf{Jac}(X) \longrightarrow T^*\mathsf{Jac}(X)$$

(3.3.7) $$\big(\lambda, (\mathsf{q}, \mathsf{p})\big) \longmapsto (\mathsf{q}, \lambda\mathsf{p})$$

which evidently extends to a holomorphic $\mathbb{C}^*$-action on $T^*\mathsf{Jac}(X)$. The fixed point set of each action equals the zero section $\mathsf{p} = 0$, which is precisely the critical set of e. Furthermore the $\mathbb{C}^*$-action extends to an action of the multiplicative semigroup $\mathbb{C}$; the limit as $\lambda \longrightarrow 0$ lies in the critical set. We revisit this geometry in §6.8.

4. Equivalence of de Rham and Dolbeault groupoids

The Betti, de Rham, and Dolbeault deformation theories of the previous sections are equivalent. Hence the three moduli *sets* are isomorphic. However these moduli sets have natural differentiable structures (in fact they are complex Lie groups). With this extra structure, we refer to the moduli sets as *moduli spaces*. The equivalence between the de Rham groupoid and the Dolbeault groupoid induces an isomorphism of real Lie groups between their moduli spaces.

4.1. Construction of the equivalence. Passage from de Rham to Dolbeault groupoids requires the construction of a holomorphic structure and a Higgs field for each flat connection D. To this end, we introduce a Hermitian metric H on E to decompose D into a unitary connection and a 1-form ϕ. The unitary connection determines a

unique holomorphic structure D''. The condition that ϕ determines a Higgs field is equivalent to a *harmonicity* condition on H, which follows from the standard theory of harmonic forms.

Hermitian metrics. A *Hermitian metric* H on E is a family of positive definite Hermitian forms $\langle,\rangle_H$

$$E_x \times E_x \longrightarrow \mathbb{C}$$

on the fibers E_x smoothly varying with $x \in \Sigma$. For any vector space V, let $\mathsf{Her}(V)$ denote the space of positive definite Hermitian forms $V \times V \longrightarrow \mathbb{C}$. The space $\mathsf{Her}(E)$ of Hermitian metrics on the vector bundle E is the space of sections of the $\mathsf{Her}(\mathbb{C}^r)$-bundle associated to E. In terms of a basis, a Hermitian form is represented by a Hermitian matrix h:

$$H(u,v) = u^\dagger h \bar{v}$$

where † indicates transpose. The action of a linear transformation $g \in \mathrm{GL}(r,\mathbb{C})$ on a Hermitian form defined by a matrix h is then:

$$g : h \longmapsto g^\dagger h \bar{g}.$$

If E is a flat vector bundle with holonomy $\phi : \pi \longrightarrow \mathrm{GL}(r,\mathbb{C})$, then a Hermitian metric $H \in \mathsf{Her}(E)$ corresponds to a ϕ-equivariant map

(4.1.1) $$\tilde{\Sigma} \xrightarrow{h} \mathsf{Her}(\mathbb{C}^r).$$

In particular $\mathsf{End}(E)$ is a complex line bundle with a canonical *everywhere nonzero* section, the identity endomorphism $\mathbb{I} : E_x \longrightarrow E_x$. Hence $\mathsf{End}(E)$ identifies canonically with the trivial line bundle $\mathbb{C}$ over X.

The associated bundles of Hermitian metrics are trivial, but not canonically trivial.

However, in the important special case when the holonomy lies in the subgroup $\mathrm{U}(1)$, the standard Hermitian form on $\mathbb{C}$

$$\langle z, w \rangle := z\bar{w}$$

defines a Hermitian metric on E. Gauge transformations act on Hermitian metrics by:

$$\langle u, v \rangle_{\xi \cdot H} := \langle g^{-1} \cdot u, g^{-1} \cdot v \rangle_H$$

and, in rank one, the corresponding positive function h transforms by:

$$h \longmapsto |g|^{-2} h.$$

Two Hermitian metrics $H_1, H_2 \in \mathsf{Her}(E)$ on a complex line bundle E relate by their *ratio*, the positive function $h : X \longrightarrow \mathbb{R}^+$ defined by:

$$\langle u, v \rangle_{H_1} = h(x) \langle u, v \rangle_{H_2}$$

where $u, v \in E_x$.

LEMMA 4.1.1. *The linear gauge group $\mathcal{G}_l(E)$ acts transitively on the space $\mathsf{Her}(E)$ of Hermitian metrics.*

PROOF. If h_1, h_2 are positive functions representing Hermitian metrics on E, then
$$g(z) = \sqrt{h_1(z)/h_2(z)}$$
defines a gauge transformation $g \in \mathcal{G}_l(E)$ with $g \cdot h_2 = h_1$. □

The unique Hermitian metric H_0 for which
$$\langle \tau, \tau \rangle_{H_0} = 1$$
corresponds to the constant function 1 and its stabilizer in $\mathcal{G}_l(E)$ is the *unitary gauge group* $\mathcal{G}_u(E)$, corresponding to functions $X \longrightarrow \mathrm{U}(1)$.

A Hermitian metric H on E induces Hermitian pairings over $\mathcal{A}^*(X)$
$$\mathcal{A}^k(X; E) \times \mathcal{A}^l(X; E) \longrightarrow \mathcal{A}^{k+l}(X).$$

A connection D is *unitary* with respect to a metric H if and only if
$$d\langle s_1, s_2 \rangle_H = \langle Ds_1, s_2 \rangle_H + \langle s_1, Ds_2 \rangle_H,$$
for sections $s_1, s_2 \in \mathcal{A}^0(X, E)$. Equivalently we say that H is *parallel* with respect to D. With respect to τ_0, the connection $D = D_0 + \eta$ is unitary with respect to H if and only if:

(4.1.2) $$\eta + \bar{\eta} = 2\operatorname{Re}(\eta) = h^{-1}dh = d\log h.$$

Thus $D = D_0 + \eta \in \mathcal{F}_l(E)$ is unitary with respect to some Hermitian metric on E if and only if the 1-form $\operatorname{Re}(\eta)$ is exact.

In particular a connection $D = D_0 + \eta$ is unitary with respect to the Hermitian metric H_0 corresponding to the trivialization τ_0 (that is, the Hermitian metric with $h \equiv 1$) if and only if $\eta \in \mathcal{A}^1(X; i\mathbb{R})$. We denote the subset of $\mathcal{F}_l(E)$ consisting of flat connections unitary with respect to H_0 by $\mathcal{F}_u(E)$. Since $\mathcal{G}_l(E)$ acts transitively on $\mathsf{Her}(E)$, a connection $D \in \mathcal{F}_l(E)$ is $\mathcal{G}_l(E)$-equivalent to a connection in $\mathcal{F}_u(E)$ if and only if D is unitary with respect to some Hermitian metric on E.

The following well known result relates holomorphic structures, Hermitian structures and unitary connections:

PROPOSITION 4.1.2. *Let E be a complex line bundle over X with Hermitian metric H, and let D'' be a holomorphic structure on E. Then there exists a unique connection D on E such that*
- *The $(0,1)$-part $D^{0,1}$ of D equals D'';*
- *H is parallel with respect to D.*

We call D the connection *compatible* with D'' and H.

PROOF. Write $D'' = D_0'' + \mu$, where $\mu \in \mathcal{A}^{0,1}(X)$. Let h be the equivariant positive function on $\tilde{X}$ corresponding to H as in (4.1.1). Let $D = D_0 + \eta$ where
$$\eta = \mu - \bar{\mu} + h^{-1}\partial h.$$
Since $\eta + \bar{\eta} = h^{-1}dh$, (4.1.2) implies that D is unitary with respect to H. Since $\eta^{0,1} = \mu$, the holomorphic structure defined by D equals D'' as desired.

For uniqueness, let D_1, D_2 be two U(1)-connections with the same $(0,1)$-part. By Lemma 4.1.1, we may assume $H = H_0$. With respect to the trivialization τ_0, write $D_i = D_0 + \psi_i$ where D_0 is the trivial connection. Then ψ_1 and ψ_2 are 1-forms with values in $i\mathbb{R}$. Their difference $\psi_1 - \psi_2$ has values in $i\mathbb{R}$ with Hodge type (1,0). Since a purely imaginary 1-form of Hodge type (1,0) is necessarily zero, $\psi_1 - \psi_2$ is zero. $\square$

(Although we only proved Lemma 4.1.1 and Proposition 4.1.2 for line bundles, these results hold for vector bundles of arbitrary rank.)

The unitary case. Suppose D is a connection. Its composition
$$\mathcal{A}^0(X;E) \xrightarrow{D} \mathcal{A}^1(X;E) \longrightarrow \mathcal{A}^{0,1}(X;E)$$
with $(0,1)$-projection is a holomorphic structure, denoted $D^{0,1}$. If
$$D = D_0 + \eta,$$
then the corresponding holomorphic structure is
$$D^{0,1} = D_0'' + \Psi$$
where $\Psi = \eta^{0,1}$.

The correspondence between holomorphic structures on E and flat unitary connections is almost an equivalence of deformation theories. Define a functor
$$\big(\mathcal{F}_u(E), \mathcal{G}_u(E)\big) \xrightarrow{\mathcal{S}} \big(\mathsf{Hol}(E), \mathcal{G}_l(E)\big),$$
which on objects is the map assigning to a flat U(1)-connection D on E the holomorphic structure $D^{0,1}$, and on morphisms is the inclusion $\mathcal{G}_u(E) \hookrightarrow \mathcal{G}_l(E)$.

Unfortunately $\mathcal{S}$ is not full. Consider the connection D_0 on E and D_0'' the trivial holomorphic structure. Then
$$\mathsf{Mor}(D_0, D_0) \longrightarrow \mathsf{Mor}(D_0'', D_0'')$$
is not surjective: $\mathsf{Mor}(D_0, D_0)$ consists of scalar multiplication by unit complex numbers, while $\mathsf{Mor}(D_0'', D_0'')$ corresponds to scalar multiplication by $\mathbb{C}^*$.

4. EQUIVALENCE OF DE RHAM AND DOLBEAULT GROUPOIDS

To overcome this difficulty, rigidify the groupoid $(\mathsf{Hol}(E), \mathcal{G}_l(E))$ by including a Hermitian metric as an extra piece of data. Therefore $\mathcal{G}_l(E)$ acts on $\mathsf{Hol}(E) \times \mathsf{Her}(E)$.

Suppose $D'' \in \mathsf{Hol}(E)$ and $H \in \mathsf{Her}(E)$. Let D be the unique connection compatible with D'' and H (Proposition 4.1.2).

DEFINITION 4.1.3. *A Hermitian metric H is* adapted *to D'' if and only if the connection D compatible with D'' and H is flat, that is, $F(D) = 0$.*

(In Corlette [**C2**], such metrics are called *harmonic*, with respect to a holomorphic line bundle. To avoid confusion, we chose to restrict the terminology "harmonic metric" to the context of flat bundles and refer to the corresponding notion for holomorphic bundles as "adapted metric".)

Let $\mathsf{Hol}_u(E)$ be the subset of $\mathsf{Hol}(E) \times \mathsf{Her}(E)$ consisting of all (D'', H) such that H is adapted to D'' in the sense of Definition 4.1.3. Then $\mathcal{G}_l(E)$ acts on $\mathsf{Hol}_u(E)$, and we denote the corresponding deformation theory $(\mathsf{Hol}_u(E), \mathcal{G}_l(E))$.

PROPOSITION 4.1.4. *The natural map*

$$\mathcal{F}_u(E) \longrightarrow \mathsf{Hol}_u(E)$$
$$D \longmapsto (D^{0,1}, 1)$$

is equivariant with respect to the natural $\mathcal{G}_l(E)$-action. The corresponding functor

$$(\mathcal{F}_u(E), \mathcal{G}_u(E)) \xrightarrow{\mathcal{T}} (\mathsf{Hol}_u(E), \mathcal{G}_l(E))$$

is an equivalence of deformation theories.

PROOF. Let $(D'', h) \in \mathsf{Hol}_u(E)$. By applying a linear gauge transformation, we may assume that $h = 1$. Apply Proposition 4.1.2, to obtain a U(1)-connection D satisfies $D^{0,1} = D''$. By definition of $\mathsf{Hol}_u(E)$, the connection D is unitary, whence $D \in \mathcal{F}_u(E)$. Thus $\mathcal{T}_*$ is surjective.

Since the subgroup of $\mathcal{G}_l(E)$ preserving H_0 is $\mathcal{G}_u(E)$, the induced map

$$\mathrm{Mor}(D_1, D_2) \xrightarrow{\mathcal{T}(D_1, D_2)} \mathrm{Mor}(D_1'', D_2'')$$

is an isomorphism. Thus $\mathcal{T}$ is faithful and full. □

PROPOSITION 4.1.5. *For each $D'' \in \mathsf{Hol}(E)$, there exists $h \in \mathsf{Her}(E)$ such that $(D'', h) \in \mathsf{Hol}_u(E)$.*

PROOF. Let $D'' \in \mathsf{Hol}(E)$. One must find a flat unitary connection which is equivalent by a gauge transformation in $\mathcal{G}_l(E)$ to a connection

whose $(0,1)$-part equals D''. Write $D'' = D_0'' + \Psi$ where $\Psi \in \mathcal{A}^{0,1}(X)$. Now
$$\mathcal{A}^{0,1}(X) = \mathcal{H}^{0,1}(X) \oplus \overline{\partial}\mathcal{A}^0(X)$$
and $\Psi = \Psi_0 + \overline{\partial}s$, where Ψ_0 is antiholomorphic. (§3). In particular Ψ_0 is closed. Let
$$D = D_0 + \Psi_0 - \bar{\Psi}_0$$
and $g = \exp(s)$. Then D is flat (since $d\Psi_0 = d\bar{\Psi}_0 = 0$), unitary with respect to 1 (since $\Psi_0 - \bar{\Psi}_0$ is purely imaginary). Furthermore the gauge transformation g takes
$$D^{0,1} = D_0'' + \Psi_0$$
to D''. The metric H_0 corresponds to the function $\bar{g}g$. The desired adapted metric is $g \cdot H_0$. $\square$

Together, Propositions 4.1.4 and 4.1.5 imply:

COROLLARY 4.1.6. *The induced map*
$$\mathcal{F}_u(E)/\mathcal{G}_u(E) \longrightarrow \mathsf{Hol}(E)/\mathcal{G}_l(E)$$
is an isomorphism of sets.

That every topologically trivial holomorphic line bundle over X arises from a unitary character of $\pi_1(X)$ is a standard result; see Weyl[**Wey**], §18, Gunning [**Gu1**], §8a, pp. 131–135, or Farkas-Kra [**FK**], §III.9, pp. 119–129. In the terminology of [**FK**], a *character* is a homomorphism $\pi_1(X) \longrightarrow \mathbb{C}^*$ and a *normalized character* is a unitary character. A *multiplicative function* in the sense of [**Wey, FK**] corresponds to a holomorphic section of a flat complex line bundle.

The linear case. Now we extend the preceding theory from U(1)-representations and flat unitary connections to $\mathbb{C}^*$-representations and flat linear connections.

Suppose $D \in \mathcal{F}_l(E)$. With respect to τ_0, write $D = D_0 + \eta$, where $\eta \in \mathcal{Z}^1(X)$. Let ϕ, ψ be the real and imaginary parts of η respectively. Since D is flat,
$$d\phi + d\psi = 0.$$
Since $d\psi$ is purely imaginary and $d\phi$ is real, this single equation is equivalent to the pair of equations:
$$d\phi = d\psi = 0.$$
Thus D is flat if and only if ϕ and ψ are closed.

For any Hermitian metric H on E, decompose the connection D into its skew-adjoint part (a connection D_H compatible with H) and

4. EQUIVALENCE OF DE RHAM AND DOLBEAULT GROUPOIDS

its self-adjoint part, a 1-form ϕ_1 which is self-adjoint with respect to H:

(4.1.3) $$D = D_H + \phi_1,$$

where

$$D_H = D_0 + \psi + \frac{1}{2}h^{-1}dh$$

and

$$\phi_1 = \phi - \frac{1}{2}h^{-1}dh$$

where h is the positive function corresponding to H as in §4.1.

Then D_H is flat ($h^{-1}dh$ is exact) and compatible with H (by (4.1.2)). Since D_H is flat, ϕ_1 is a closed 1-form. Let $\Phi = (\phi_1)^{1,0}$.

Harmonic metrics. We also recall basic facts about harmonic functions. A smooth function f is *harmonic* if and only if $d \star df = 0$. In particular holomorphic functions are harmonic. A harmonic 1-form locally is the differential of a harmonic function. Every harmonic function is locally the real part of a holomorphic function. A smooth function $f : X \longrightarrow \mathbb{R}^+$ is *multiplicatively harmonic* if and only if its logarithm

$$\log f : \tilde{X} \longrightarrow \mathbb{R}$$

is a harmonic function.

DEFINITION 4.1.7. *Let D be a flat connection on the line bundle E over X. A Hermitian metric $H \in \mathsf{Her}(E)$ is harmonic with respect to D if and only if the equivariant map*

$$\tilde{X} \xrightarrow{\tilde{h}} \mathbb{R}^+$$

corresponding to H is a multiplicatively harmonic function on $\tilde{X}$, that is, its logarithm $\log \tilde{h}$ is harmonic function $X \longrightarrow \mathbb{R}$.

This definition is independent of the trivialization of D over the universal covering space $\tilde{X} \longrightarrow X$ used to define $\tilde{h}$.

H is a harmonic metric with respect to D if and only if the 1-form $\phi_1 = \phi - \frac{1}{2}h^{-1}dh$ is harmonic, or equivalently $\overline{\partial}\Phi = 0$.

LEMMA 4.1.8. *Let D be a flat connection. Then there exists a Hermitian metric H harmonic with respect to D. Furthermore H is unique up to multiplication by a positive constant.*

PROOF. Write $D = D_0 + \psi + \phi$ as above, where $\phi \in \mathcal{Z}^1(X, \mathbb{R})$. Then ϕ is cohomologous to a unique harmonic form $\phi_1 \in \mathcal{H}^1(X; \mathbb{R})$, that is,

$$\phi = \phi_1 + df.$$

for some smooth function $f \in \mathcal{A}^0(X, \mathbb{R})$ (§3). The desired harmonic metric is $h = e^{2f}$. Since f is unique up to addition of a real constant, the metric H is unique up to multiplication by a positive constant. □

In higher rank, these facts generalize to deeper technical results. The existence and uniqueness of flat unitary connections on a topologically trivial holomorphic line bundle (Proposition 4.1.5) generalizes to Donaldson's proof of the Narasimhan-Seshadri theorem [**D1**], Hitchin's proof of solutions of the self-duality equations in rank two [**H1**] and Simpson's proof in general [**Si1, Si2**]. Lemma 4.1.8 generalizes to the existence of harmonic metrics on flat bundles with reductive holonomy (Donaldson [**D2**] in rank two, and Corlette [**C1**] in general). The existence of a harmonic metric generalizes the existence and uniqueness of harmonic maps of compact Riemannian manifolds into complete Riemannian manifolds of nonnegative curvature, due to Eells-Sampson [**ES**], and has been further generalized by Labourie [**L**]).

Let $D \in \mathcal{F}_l(E)$ and H its harmonic metric and $D = D_H + \phi_1$, where D_H is compatible with H. Let D'' be the (0,1)-part of D_H and Φ be the (1,0)-part of ϕ_1. Since ϕ_1 is harmonic, $\bar{\partial}\Phi = 0$. Thus (E, D'', Φ) is a Higgs pair.

THEOREM 4.1.9. *The resulting functor*

$$(\mathcal{F}_l(E), \mathcal{G}_l(E)) \xrightarrow{\mathcal{S}} (\mathsf{Higgs}(E), \mathcal{G}_l(E)).$$

is an equivalence of deformation theories.

PROOF. First we show that $\mathcal{S}$ is surjective on isomorphism classes. Let $(D'', \Phi) \in \mathsf{Higgs}(E)$. By Proposition 4.1.5, D'' has an adapted metric H corresponding to a function $h : X \to \mathbb{R}^+$. Let D_H be the unique connection compatible with H and $(D_H)^{0,1} = D''$ (Proposition 4.1.2). Write $D_H = D_0 + \psi$. Then $\xi \cdot H = H_0$ for some linear gauge transformation $\xi \in \mathcal{G}_l(E)$. Let

$$D = \xi \cdot (D_H + \Phi + \bar{\Phi}) = D_0 + \Phi + \bar{\Phi}.$$

The curvature of D is

$$F(D) = F(\xi \cdot D_H) + d(\Phi + \bar{\Phi}).$$

Since H is adapted to D'', the curvature $F(D_H) = 0$, so $F(\xi \cdot D_H) = 0$. Since Φ is holomorphic,

$$d(\Phi + \bar{\Phi}) = 0.$$

Hence $F(D) = 0$ and $D \in \mathcal{F}_l(E)$. Then $\mathcal{S}(D)$ is a Higgs pair differing from (D'', Φ) by the gauge transformation ξ.

4. EQUIVALENCE OF DE RHAM AND DOLBEAULT GROUPOIDS

Next we show $\mathcal{S}$ is faithful and full. Suppose
$$(D_1'', \Phi_1), (D_2'', \Phi_2) \in \mathsf{Higgs}(E).$$
There are two cases, depending on whether or not $(D_1'', \Phi_1), (D_2'', \Phi_2)$ are $\mathcal{G}_l(E)$-equivalent. If $(D_1'', \Phi_1), (D_2'', \Phi_2)$ are not $\mathcal{G}_l(E)$-equivalent, then
$$\mathsf{Mor}\big((D_1'', \Phi_1), (D_2'', \Phi_2)\big)$$
is empty. Otherwise, there exists $\xi \in \mathcal{G}_l(E)$ corresponding to a map $g \in \mathsf{Map}(X, \mathbb{C}^*)$ such that
$$D_2'' = D_1'' + g^{-1}\bar{\partial}g, \quad \Phi_2 = \Phi_1$$
and if ξ_1 is another gauge transformation corresponding to a map $g_1 \in \mathsf{Map}(X, \mathbb{C}^*)$ such that
$$D_2'' = D_1'' + g_1^{-1}\bar{\partial}g_1,$$
then $g_1 = gc$, where c is a constant map. We denote the subgroup of $\mathcal{G}_l(E)$ corresponding to constant maps $X \longrightarrow \mathbb{C}^*$ by $\mathbb{C}^*$. Hence $\mathsf{Mor}\big((D_1'', \Phi_1), (D_2'', \Phi_2)\big)$ corresponds to the coset $g \cdot \mathbb{C}^*$.

Now suppose that $D_1, D_2 \in \mathcal{F}_l(E)$ and $\mathcal{S}(D_i) = (D_i'', \Phi_i)$ for $i = 1, 2$. If D_1 is not $\mathcal{G}_l(E)$-equivalent to D_2, then then $\mathsf{Mor}(D_1, D_2) = \emptyset$ since $\mathbb{C}^*$ acts trivially on $\mathcal{F}_l(E)$. Otherwise, $D_1 = g \cdot D_2$ where $g \in \mathcal{G}_l(E)$. Then $\mathsf{Mor}(D_1, D_2) = g \cdot \mathbb{C}^*$.

In both cases, $\mathcal{S}$ induces an isomorphism
$$\mathsf{Mor}(D_1, D_2) \longrightarrow \mathsf{Mor}(\mathcal{S}(D_1), \mathcal{S}(D_2))$$
as desired. $\square$

COROLLARY 4.1.10. *The induced map*
$$\mathcal{F}_l(E)/\mathcal{G}_l(E) \xrightarrow{\mathcal{S}_*} \mathsf{Higgs}(E)/\mathcal{G}_l(E)$$
is an isomorphism of sets.

4.2. Higgs coordinates. To see that the bijection $\mathcal{S}_*$ between the de Rham and Dolbeault moduli spaces is an isomorphism of real Lie groups, we examine in detail the constructions in the proof of Theorem 4.1.9. We explicitly describe how to pass from the equivalence class of a flat connection $D = D_0 + \eta$ to an equivalence class of a Higgs pair (D'', Φ), where $D'' = D_0'' + \Psi$. We call $\eta \in \mathcal{H}^1(X)$ the *connection coordinates* of the point in $\mathcal{F}_l(E)/\mathcal{G}_l(E)$ and $(\Psi, \Phi) \in V \oplus \bar{V}$ the *Higgs coordinates*.

Passing from connection coordinates $\eta \in \mathcal{H}(X)$ to Higgs coordinates $(\Psi, \Phi) \in V \oplus \bar{V}$ involves the following steps.

We assume that η is harmonic, since we are only interested in the equivalence class of D under $\mathcal{G}_l(E)$. Decompose η into its imaginary part ψ and its real part ϕ:

$$\eta = \psi + \phi$$

where

$$\psi = \frac{1}{2}(\eta - \bar{\eta}), \qquad \phi = \frac{1}{2}(\eta + \bar{\eta}).$$

The unitary connection $D_H := D_0 + \psi$ corresponds to a holomorphic structure $D'' = D_0'' + \Psi$ where the antiholomorphic 1-form Ψ is the $(0,1)$-part of ψ:

$$\Psi = \psi^{0,1}$$
$$\psi = \Psi - \bar{\Psi}.$$

The $(1,0)$-part Φ of ϕ is a holomorphic 1-form:

$$\Phi = \phi^{1,0}$$
$$\phi = \Phi + \bar{\Phi}.$$

In summary, to pass from Higgs coordinates to connection coordinates:

(4.2.1) $$\eta = \Psi - \bar{\Psi} + \Phi + \bar{\Phi}$$

and to pass from connection coordinates to Higgs coordinates:

(4.2.2) $$\Psi = (i \operatorname{Im} \eta)^{0,1}$$
$$\Phi = (\operatorname{Re} \eta)^{1,0}$$

Holomorphic structures associated to flat connections. Since

$$\eta = \psi + \phi$$
$$= (\Psi - \bar{\Psi}) + (\Phi + \bar{\Phi})$$
$$= (\Psi + \bar{\Phi}) + (\Phi - \bar{\Psi})$$

where $\Psi \in \mathcal{H}^{0,1}(X)$ and $\Phi \in \mathcal{H}^{1,0}(X)$, the Hodge components of η are:

$$\eta^{0,1} = \Psi + \bar{\Phi}$$
$$\eta^{1,0} = \Phi - \bar{\Psi}.$$

The flat connection D determines a holomorphic structure:

$$D^{0,1} := D_0'' + \eta^{0,1} = D_0'' + \Psi + \bar{\Phi} = D'' + \bar{\Phi}.$$

The holomorphic structure D'' in the Higgs pair (D'', Φ) is:

$$D'' := D_0'' + \Psi.$$

Thus, unless $\Phi = 0$, the holomorphic structure in the Higgs pair is *not* the holomorphic structure determined by the flat connection.

4. EQUIVALENCE OF DE RHAM AND DOLBEAULT GROUPOIDS

Isomorphism of Lie groups. Now we strengthen Corollary 4.1.10:

PROPOSITION 4.2.1. *The induced map*
$$\mathcal{F}_l(E)/\mathcal{G}_l(E) \xrightarrow{\mathcal{S}_*} \mathsf{Higgs}(E)/\mathcal{G}_l(E)$$
is an isomorphism of real Lie groups

PROOF. Using harmonic decomposition, the moduli space $\mathcal{F}_l(E)/\mathcal{G}_l(E)$ identifies with the quotient of $\mathcal{H}^1(X)$ by the discrete subgroup corresponding to the subgroup $\mathrm{H}^1(X,\mathbb{Z}) \hookrightarrow \mathrm{H}^1(X)$. The moduli space $\mathsf{Higgs}(E)/\mathcal{G}_l(E)$ corresponds to the quotient of $V \oplus \bar{V}$ by a lattice $L \subset V$ in the first factor. The map $\mathcal{S}_*$ arises from the $\mathbb{R}$-linear map
$$\mathcal{H}^1(X) \longrightarrow V \oplus \bar{V}$$
$$\eta \longmapsto (\Psi, \Phi)$$
in Higgs coordinates. The induced bijection $\mathcal{S}_*$ is an isomorphism of real Lie groups. □

Real structures. The anti-involutions defining the real forms $\mathrm{U}(1)$ and $\mathbb{R}^*$ were described in connection coordinates in (2.2.8). Using (4.2.1) and (4.2.2) to pass to Higgs coordinates, $(\iota_U)_*$ and $(\iota_\mathbb{R})_*$ are:

(4.2.3)
$$(\Psi, \Phi) \xmapsto{(\iota_U)_*} (\Psi, -\Phi)$$
$$(\Psi, \Phi) \xmapsto{(\iota_\mathbb{R})_*} (-\Psi, \Phi).$$

The fixed point set of $(\iota_U)_*$ equals $\mathcal{F}_u(E)/\mathcal{G}_u(E)$. In the description of the moduli space as $T^*\mathsf{Jac}(X)$, the involution ι_U is scalar multiplication by -1 on the fibers of $T^*\mathsf{Jac}(X) \longrightarrow \mathsf{Jac}(X)$.

In contrast, $\iota_\mathbb{R}$ leaves Φ unchanged, but replaces the holomorphic structure D'' by its inverse. In the cotangent bundle description, $\iota_\mathbb{R}$ is the lift of inversion
$$\mathsf{Jac}(X) \longrightarrow \mathsf{Jac}(X)$$
$$L \longmapsto L^{-1}$$
to $T^*\mathsf{Jac}(X)$ which fixes the parallelism
$$T^*\mathsf{Jac}(X) \longrightarrow T^*_{L_0}\mathsf{Jac}(X) \cong V^* \cong \bar{V}.$$

The differential of inversion induces the composition of this map with the map ι_U above given by scalar multiplication by -1 on the cotangent fibers.

The fixed points of inversion form the 2-torsion subgroup
$$\mathsf{Jac}_2(X) \hookrightarrow \mathsf{Jac}(X)$$

corresponding to
$$\left(\frac{1}{2}L\right)/L \subset V/L$$
which is isomorphic to $(\mathbb{Z}/2)^{2k}$ where k is the genus of X. Thus the fixed point set of $\iota_\mathbb{R}$ acting on $\mathsf{Hom}(\pi, \mathbb{C}^*)$ is the subset $\mathsf{Hom}(\pi, \mathbb{R}^*)$ of
$$\mathsf{Hom}(\pi, \mathbb{C}^*) \longleftrightarrow T^*\mathsf{Jac}(X)$$
corresponding to
$$T^*_{\mathsf{Jac}(X)}\big|_{\mathsf{Jac}_2(X)} \cong (\mathbb{Z}/2)^{2k} \times \bar{V}.$$

4.3. Involutions. The involution ι_U is holomorphic with respect to I, but anti-holomorphic with respect to J. The fixed point set of $t = -1$ in the $\mathbb{C}^*$-action is holomorphic with respect to I, but is a purely real subset with respect to J. Hence this fixed point set is $\mathsf{Hom}(\pi, \mathrm{U}(1))$ which embeds into $\mathsf{Higgs}(E)/\mathcal{G}_l(E)$ as a holomorphic submanifold by (4.2.3).

Similarly, the involution $\iota_\mathbb{R}$ is holomorphic with respect to I and antiholomorphic with respect to J. The fixed point set of $\iota_\mathbb{R}$ is $\mathsf{Hom}(\pi, \mathbb{R}^*)$ which is holomorphic with respect to I.

The common fixed point set of ι_U and $\iota_\mathbb{R}$ is the intersection
$$\mathsf{Hom}(\pi, \mathrm{U}(1)) \cap \mathsf{Hom}(\pi, \mathbb{R}^*) = \mathsf{Hom}(\pi, \pm\mathbb{I})$$
which identifies with the 2-torsion subgroup $\mathsf{Jac}_2(X)$ of the Jacobian. (This is also the fixed point set of the composition $\iota_U \circ \iota_\mathbb{R}$.) In particular $\mathsf{Hom}(\pi, \mathbb{R}^*)$ corresponds to Higgs pairs (D'', Φ) where the holomorphic line bundle (E, D'') has order two in the Jacobian. Its identity component $\mathsf{Hom}(\pi, \mathbb{R}^+)$ corresponds to Higgs bundles whose underlying holomorphic structure is trivial.

The holomorphic structure of the Higgs pair corresponding to an $\mathbb{R}^+$-representation is trivial. If $\rho: \pi \longrightarrow \mathbb{R}^+$ is a nontrivial real character, then there exists a global holomorphic section to the corresponding flat line bundle but *no* global parallel section. This holomorphic section arises as follows. Under the identifications
$$\mathsf{Hom}(\pi, \mathbb{R}^+) \xrightarrow{\log} \mathsf{Hom}(\pi, \mathbb{R}) \cong \mathrm{H}^1(X; \mathbb{R})$$
ρ corresponds to a harmonic 1-form ϕ. On the universal cover, ϕ lifts to an exact 1-form which is the differential of a harmonic function $f: \tilde{X} \longrightarrow \mathbb{R}$ satisfying:

(4.3.1) $$f \circ \gamma - f = \log \rho(\gamma).$$

There exists a holomorphic function $F: \tilde{X} \longrightarrow \mathbb{C}$ with $\mathrm{Re}\, F = f$ which necessarily satisfies (4.3.1). (The function F is unique up to an additive

purely imaginary constant.) Then $\exp F$ is a holomorphic function on $\tilde{X}$ and defines a nonvanishing holomorphic section of the corresponding flat line bundle.

This holomorphic section is nonvanishing for the following general reason. Since every flat line bundle has degree zero, a nonzero holomorphic section has no zeroes. Compare the discussion in Gunning [**Gu2**], pp.237–238. The function f arose earlier in the construction of the harmonic metric in Lemma 4.1.8.

In general the direct product decomposition of Lie groups $\mathbb{C}^* \cong U(1) \times \mathbb{R}^+$ induces a decomposition of Betti moduli spaces

$$\mathsf{Hom}(\pi, \mathbb{C}^*) \cong \mathsf{Hom}(\pi, U(1)) \times \mathsf{Hom}(\pi, \mathbb{R}^+).$$

The corresponding decomposition of Dolbeault moduli spaces

$$\mathsf{Higgs}(E)/\mathcal{G}_l(E) \cong \mathsf{Jac}(X) \times \mathrm{H}^1(X; \mathbb{R})$$

associates to a Higgs pair the underlying holomorphic structure and the class of the logarithmic differential of the harmonic metric in $\mathrm{H}^1(X; \mathbb{R})$.

5. Hyperkähler geometry on the moduli space

Theorem 2.3.1 and Corollaries 4.1.6 and 4.1.10 establish bijections between the three moduli spaces

$$(\mathbb{C}^*)^{2k}, \quad \mathrm{H}^1(\Sigma)/\mathrm{H}^1(\Sigma, \mathbb{Z}), \quad T^*\mathsf{Jac}(X)$$

which are isomorphisms of real Lie groups. The first two — Betti and de Rham — were isomorphic as complex Lie groups with isomorphic complex-symplectic structures. We henceforth will only work with the de Rham and Dolbeault moduli spaces.

The complex structure J on the de Rham moduli space and the complex structure I on the Dolbeault space are quite different. Using the isomorphism $\mathcal{S}_*$, we superimpose these structures on a single object which we call $\mathcal{M}$. The interaction between these structures leads to a rich *hyperkähler geometry* and group actions on $\mathcal{M}$.

5.1. The quaternionic structure. In connection coordinates $\eta \in \mathcal{H}^1(X)$ and Higgs coordinates $(\Psi, \Phi) \in V \oplus \bar{V}$, the complex structure I on $\mathcal{M}$ is:

(5.1.1)
$$\eta \xmapsto{I} - \star \bar{\eta}$$
$$(\Psi, \Phi) \longmapsto (i\Psi, i\Phi).$$

The formula in connection coordinates may be derived as follows. Let $(\Psi, \Phi) \in V \oplus \bar{V}$ be the Higgs coordinates of η. By definition of I, the

Higgs coordinates of $I(\eta)$ are $(i\Psi, i\Phi)$ so (4.2.1) implies:

$$\begin{aligned} I(\eta) &= (i\Psi) - \overline{(i\Psi)} + (i\Phi) + \overline{(i\Phi)} \\ &= i\Psi + i\bar{\Psi} + i\Phi - i\bar{\Phi} \\ &= \star\Psi - \star\bar{\Psi} - \star\Phi - \star\bar{\Phi} \\ &= -\star\overline{(-\bar{\Psi} + \Psi + \bar{\Phi} + \Phi)} \\ &= -\star\bar{\eta}. \end{aligned}$$

Using the equivalence $\mathcal{S}_*$ with the de Rham moduli space, the natural complex structure on the de Rham moduli space (2.2.6) defines the complex structure J on $\mathcal{M}$:

(5.1.2)
$$\eta \xmapsto{J} i\eta$$
$$(\Psi, \Phi) \longmapsto (i\bar{\Phi}, -i\bar{\Psi}).$$

To derive the expression in Higgs coordinates, decompose η as in (4.2.1). Then:

$$\begin{aligned} J(\eta) = i\eta &= i\Psi - i\bar{\Psi} + i\Phi + i\bar{\Phi} \\ &= (i\bar{\Phi}) - \overline{(i\bar{\Phi})} + (-i\bar{\Psi}) + \overline{(-i\bar{\Psi})} \end{aligned}$$

so that the Higgs coordinates of $J(\eta)$ are $(i\bar{\Phi}, -i\bar{\Psi})$. Let $K := IJ$. In coordinates:

(5.1.3)
$$\eta \xmapsto{K} i \star \bar{\eta}$$
$$(\Psi, \Phi) \longmapsto (-\bar{\Phi}, \bar{\Psi}).$$

The complex structures I, J satisfy

$$I^2 = J^2 = -1, \qquad IJ = -JI$$

and together I, J, K define a left action of the *algebra of quaternions* $\mathbb{H}$ on $\mathcal{H}^1(X)$. Here $\mathbb{H}$ is the $\mathbb{R}$-algebra with basis $1, I, J, K$ and multiplication satisfying:

(5.1.4)
$$I^2 = J^2 = K^2 = -1$$
$$IJ = -JI = K, \quad JK = -KJ = I, \quad KI = -IK = J.$$

Furthermore every purely imaginary unit quaternion $u = u_I I + u_J J + u_K K$ (where $u_I, u_J, u_K \in \mathbb{R}$ satisfy $u_I^2 + u_J^2 + u_K^2 = 1$) satisfies $u^2 = -1$, obtaining a whole S^2 of complex structures. Opposite complex structures are related by *the antipodal map*:

(5.1.5)
$$S^2 \longrightarrow S^2$$
$$u \longmapsto -u.$$

5.2. The Riemannian metric.

The real part of the Hermitian form $\langle,\rangle$ on V in (3.3.1) and the real part of the corresponding Hermitian form on $\bar{V}$ define a Riemannian metric g on $\mathcal{M}$:

$$g\big((\Psi_1, \Phi_1), (\Psi_2, \Phi_2)\big) = \operatorname{Re}\langle \Psi_1, \Psi_2 \rangle + \operatorname{Re}\langle \Phi_1, \Phi_2 \rangle$$

(5.2.1)
$$g(\eta_1, \eta_2) = \int_X \eta_1 \wedge \star \overline{\eta_2}.$$

The Riemannian metric is invariant under the complex structures I, J, K:

$$g(\eta_1, \eta_2) = g(I\eta_1, I\eta_2) = g(J\eta_1, J\eta_2) = g(K\eta_1, K\eta_2)$$

and more generally $g(u\eta_1, u\eta_2) = g(\eta_1, \eta_2)$ for $u \in S^2$. Thus, for each $u \in S^2$, the Riemannian metric g is Hermitian with respect to u, that is, of type $(1, 1)$ with respect to u.

Each purely imaginary unit quaternion $u \in S^2$ determines a real-symplectic structure, or *Kähler form* ω_u:

(5.2.2)
$$\omega_u(\eta_1, \eta_2) := g(u\eta_1, \eta_2).$$

ω_u is the real-symplectic structure associated to a Kähler structure H for which the associated Riemannian structure is g, with respect to the complex structure u:

$$\operatorname{Im} H = \omega_u$$
$$\operatorname{Re} H = g.$$

The general definition of a hyperkähler structure is a quadruple

$$(g, I, J, K)$$

where g is a Riemannian metric and I, J, K are integrable almost complex structures satisfying the quaternion identities (5.1.4), for which g is Kählerian with respect to each of them.

A hyperkähler structure also determines complex-symplectic structures. For each fixed $u \in S^2$, there is a family of complex-symplectic structures which are holomorphic with respect to u. Namely, extend u to a positively oriented orthonormal frame (u, u_1, u_2), and define:

(5.2.3)
$$\Omega_{(u_1, u_2)} := \omega_{u_1} + i\omega_{u_2},$$

where the Kähler forms ω_{u_i} are defined in (5.2.2). Then $\Omega_{(u_1, u_2)}$ is $\mathbb{C}$-bilinear with respect to the complex structure u. Furthermore every other positively oriented orthonormal frame (u, u'_1, u'_2) is obtained by rotation:

$$u'_1 = \cos(\theta)u_1 - \sin(\theta)u_2$$
$$u'_2 = \sin(\theta)u_1 + \cos(\theta)u_2$$

for some $\theta \in \mathbb{R}$ and
$$\Omega_{(u_1', u_2')} = e^{i\theta} \Omega_{(u_1, u_2)}.$$

The unit vector u_2 is determined by u and u_1 as the cross-product $u \times u_1$. Thus these complex-symplectic structures are parametrized by orthonormal pairs (u, u_1). Since the tangent space to S^2 at $u \in S^2$ is the orthogonal complement $u^\perp \subset \mathbb{R}^3$, the complex-symplectic structures are parametrized by the *unit tangent bundle* $T_1 S^2$. Left-multiplication by u preserves $u^\perp$ and defines an integrable almost complex structure on S^2. In §6.1, we describe how this complex manifold identifies with the complex projective line $\mathbb{P}^1$ and the corresponding line bundle is the *holomorphic tangent bundle* of P^1. It has degree 2 and is denoted $\mathcal{O}(2)$. In §6 these complex-symplectic structures on the moduli space unify into one single holomorphic exterior 2-form on the twistor space which takes values in a line bundle induced from $\mathcal{O}(2)$.

5.3. Complex-symplectic structure.

An alternative approach to hyperkähler geometry is to begin with a complex-symplectic manifold and refine the complex-symplectic structure to a hyperkähler structure. Suppose that (M, J, Ω) is a complex-symplectic manifold, that is, (M, J) is a complex manifold and Ω is closed nondegenerate complex-valued exterior 2-form of type $(2,0)$ with respect to J. A *hyperkähler structure refining* (J, Ω) consists of another complex structure I which is anti-linear with respect to J and for which the symmetric form defined by

(5.3.1) $$g(\alpha, \beta) := \operatorname{Re} \Omega(\alpha, I\beta)$$

is Kählerian with respect to J and I.

This approach can be succinctly described in terms of classical Lie groups as follows. Let $\mathbb{C}^{2n}$ be the standard $2n$-dimensional complex symplectic vector space. The complex symplectic group $\operatorname{Sp}(2n, \mathbb{C})$ is a subgroup of the automorphisms of $\mathbb{C}^{2n}$. The stabilizer of the standard Hermitian structure on $\mathbb{C}^{2n}$ is the compact symplectic group

$$\operatorname{Sp}(2n) = \operatorname{Sp}(2n, \mathbb{C}) \cap \operatorname{U}(2n),$$

the maximal compact subgroup of $\operatorname{Sp}(2n, \mathbb{C})$.

A reduction of structure group for the tangent bundle of a $4n$-dimensional real manifold from $\operatorname{GL}(4n, \mathbb{R})$ to $\operatorname{Sp}(2n, \mathbb{C})$ consists of a pair (Ω, J) where J is an almost complex structure and Ω a J-bilinear complex-valued nondegenerate exterior 2-form on M. This reduction of structure group is *integrable*, that is, corresponds to a complex-symplectic structure on M if and only if J is integrable and Ω is closed.

5. HYPERKÄHLER GEOMETRY ON THE MODULI SPACE

A further reduction of structure group from $\mathrm{Sp}(2n, \mathbb{C})$ to its maximal compact subgroup $\mathrm{Sp}(2n)$ consists of another complex structure I which is J-anti-linear, and such that (5.3.1) defines a positive definite Hermitian metric with respect to I. Such a reduction of structure group gives tensor fields g, I, J, K etc. which satisfy the hyperkähler identities. For further examples and discussion of hyperkähler geometry, we refer the reader to Besse [**B**], Dancer [**D**], Hitchin [**H3**], Hitchin-Karlhede-Lindström-Roček [**HKLR**], Joyce [**Joy**], or Salamon [**Sa**].

Relation to the moduli spaces. These objects relate to the J-holomorphic complex-symplectic structure Ω on the de Rham moduli space in §2.2. By the definitions ((2.2.7), (3.2.4), (5.2.1)) of Ω, I, g,

$$g(\eta_1, \eta_2) = \mathrm{Re}\ \Omega(I\eta_1, \eta_2).$$

Thus

(5.3.2) $$\Omega = \Omega_{(-I,K)} = -\omega_I + i\omega_K$$

in the notation of (5.2.3), with real and imaginary parts discussed in (2.2.9).

Similarly the I-holomorphic complex-symplectic structure Ω_{T^*} on the Dolbeault moduli space (3.3.4) is:

(5.3.3) $$\Omega_{T^*} = 8\Omega_{J,K}$$

in the notation of (5.2.3). We indicate the proof of (5.3.3) as follows: The projections onto Hodge components are:

$$\mathcal{H}^1(X) \longrightarrow \mathcal{H}^{0,1}(X) \qquad\qquad \mathcal{H}^1(X) \longrightarrow \mathcal{H}^{0,1}(X)$$
$$\eta \longmapsto \frac{1}{2}(\eta - i \star \eta) \qquad\qquad \eta \longmapsto \frac{1}{2}(\eta + i \star \eta).$$

The $(0,1)-$ and $(1,0)-$ components of $\psi = 2i\,\mathrm{Im}(\eta)$ and $\phi = 2\,\mathrm{Re}(\eta)$ respectively, are:

$$\Psi = \frac{1 - i\star}{2} \frac{\eta - \bar{\eta}}{2} = \frac{1}{4}(\eta - \bar{\eta} - i \star \eta + i \star \bar{\eta})$$
$$\Phi = \frac{1 + i\star}{2} \frac{\eta + \bar{\eta}}{2} = \frac{1}{4}(\eta + \bar{\eta} + i \star \eta + i \star \bar{\eta}).$$

Applying (4.2.1) to (3.3.4)

$$\Omega_{T^*}(\eta_1, \eta_2) = 2\Big\{(\eta_1 \wedge \eta_2 - \bar{\eta}_1 \wedge \bar{\eta}_2)$$
$$+ (\eta_1 \wedge i \star \bar{\eta}_2 - \bar{\eta}_1 \wedge i \star \eta_2) + (-i \star \eta_1 \wedge \bar{\eta}_2 + i \star \bar{\eta}_1 \wedge \eta_2)$$
$$+ (\star\eta_1 \wedge \star\eta_2 - \star\bar{\eta}_1 \wedge \star\bar{\eta}_2)\Big\}$$

Taking real parts and imaginary parts, (5.3.3) follows. The Kähler form on $T^*\mathsf{Jac}(X)$ is ω_I.

5.4. Quaternionization. The hyperkähler structure identifies the vector space $V \oplus \bar{V}$ with the tensor product (over $\mathbb{C}$)

$$V_{\mathbb{H}} := \mathbb{H} \otimes_{\mathbb{C}} V.$$

Here the quaternion algebra $\mathbb{H}$ is a complex vector space where the action of i is left-multiplication by $I \in \mathbb{H}$. This quaternionic action extends to $V_{\mathbb{H}}$, making $V_{\mathbb{H}}$ a left $\mathbb{H}$-module.

Identify $V \oplus \bar{V}$ with $V_{\mathbb{H}}$ as follows: Let $a : V \longrightarrow \bar{V}$ be the anti-isomorphism given by the identity map on $V_{\mathbb{R}}$. The $\mathbb{R}$-linear automorphism

$$(\Psi, \Phi) \xmapsto{J} (-a^{-1}(\Phi), a(\Psi))$$

of $V \oplus \bar{V}$ is anti-linear and satisfies $J^2 = -1$. The given complex structures I and J define a left $\mathbb{H}$-action on $V \oplus \bar{V}$. The resulting map

$$V_{\mathbb{H}} \longrightarrow V \oplus \bar{V}$$

is an isomorphism of left $\mathbb{H}$-modules.

The two complex structures $\pm I$ are special, since the $\mathbb{R}$-span of L is invariant under each of these structures. For those two complex structures, $V_{\mathbb{H}}/L$ contains the complex torus V/L as a compact holomorphic submanifold. However, for all other complex structures $u \in S^2$, the resulting complex manifold is *Stein:*

THEOREM 5.4.1. *Let $u \in S^2$. Suppose that $u \neq \pm I$. Then the complex manifold $(V_{\mathbb{H}}/L, \mathbb{J}_u)$ is biholomorphic to $(\mathbb{C}^*)^{2k}$ where $k = \dim_{\mathbb{C}}(V)$.*

PROOF. The lattice L spans V over $\mathbb{R}$, and V is I-invariant. Denote by $\mathbb{C}_u$ the subalgebra $\mathbb{R}1 + \mathbb{R}u$ generated by u. Since $u \neq \pm I$, the two unit quaternions I and u generate $\mathbb{H}$. Thus V generates $V_{\mathbb{H}}$ over the subalgebra $\mathbb{C}_u \subset \mathbb{H}$. Thus the $\mathbb{R}$-linear map

$$\mathbb{C} \otimes_{\mathbb{R}} V \longrightarrow V_{\mathbb{H}}$$
$$(x + yi) \otimes v \longmapsto (x + yu)v$$

is surjective. It is an isomorphism because

$$\dim_{\mathbb{R}} \left(\mathbb{C} \otimes_{\mathbb{R}} V \right) = 4k = \dim_{\mathbb{R}} V_{\mathbb{H}}.$$

As abelian groups, $V = L \otimes \mathbb{R}$, so

$$V_{\mathbb{H}} \cong \mathbb{C} \otimes_{\mathbb{R}} V \cong \mathbb{C} \otimes L$$

Since L is free abelian of rank $2k$,

$$\begin{aligned}
V_{\mathbb{H}}/L &\cong (\mathbb{C} \otimes L)/L \\
&\cong (\mathbb{C} \otimes L)/(\mathbb{Z} \otimes L) \\
&\cong (\mathbb{C}/\mathbb{Z}) \otimes L \\
&\cong (\mathbb{C}^*) \otimes L \cong (\mathbb{C}^*)^{2k}
\end{aligned}$$

as desired $\square$

6. The twistor space

A hyperkähler structure is equivalent to a holomorphic construction which embodies all the complex and complex-symplectic structures, due to Hitchin-Karlhede-Lindström-Roček [**HKLR**].

For each unit purely imaginary quaternion $u \in S^2$, left-multiplication by u describes a complex structure on $V_{\mathbb{H}}/L$. The collection of complex manifolds $(V_{\mathbb{H}}/L, u)$ unifies in a single holomorphic fibration Π_Z over S^2, where S^2 identifies with the complex projective line $\mathbb{P}^1$ by stereographic projection (6.1.1). The corresponding holomorphic fibration $Z \xrightarrow{\Pi_Z} \mathbb{P}^1$ is *the twistor space* of $\mathcal{M}$.

In this section we explicitly describe this fibration and use it to describe Simpson's $\mathbb{C}^*$-action on $\mathcal{M}$. We realize $\mathbb{P}^1$ as a space of complex structures on $\mathbb{H}$ by stereographic projection. The twistor space results, as well as a closed holomorphic exterior 2-form $\boldsymbol{\Omega}$ on Z. This exterior 2-form defines the holomorphic complex-symplectic structure on the fibers of Π_Z, taking values in the pullback $\Pi_Z^* \mathcal{O}(2)$ of the line bundle $\mathcal{O}(2)$ on $\mathbb{P}^1$. Differentiably, Z is the product bundle $V_{\mathbb{H}}/L \times \mathbb{P}^1$, but holomorphically it is nontrivial. The rational curves $l_\mathsf{v} = \{[\mathsf{v}]\} \times \mathbb{P}^1$ define a family of holomorphic sections of Z, *the twistor lines,* whose normal bundles are isomorphic to $\mathcal{O}(1)$. Finally the antipodal map $\boldsymbol{\alpha}$ on $\mathbb{P}^1$ lifts to an anti-holomorphic involution on Z which satisfies compatibility relations with the twistor lines and $\boldsymbol{\Omega}$. The basic result (Theorem 3.3 of [**HKLR**]) is that the collection $(\Pi_Z, \{l_\mathsf{v}\}, \boldsymbol{\Omega}, \widetilde{\boldsymbol{\alpha}})$ is equivalent to a hyperkähler structure. We describe these objects explicitly in this special case (Theorem 6.7.1).

Finally the $\mathbb{C}^*$-action is described in terms of the twistor construction.

6.1. The complex projective line. We briefly review the geometry of $\mathbb{P}^1$. For more details, see §1–§2 of Hitchin's article in [**HSW**].

Points in $\mathbb{P}^1$ correspond to complex lines (1-dimensional linear subspaces) in $\mathbb{C}^2$. Cover $\mathbb{P}^1$ by two affine patches $\mathbb{A}_1, \mathbb{A}_2$ with coordinate

charts:

$$\mathbb{C} \xrightarrow{\psi^{(1)}} \mathbb{A}_1 \qquad\qquad \mathbb{C} \xrightarrow{\psi^{(2)}} \mathbb{A}_2$$

$$\zeta \longmapsto \mathbb{C}\begin{bmatrix}\zeta\\1\end{bmatrix} \subset \mathbb{C}^2 \qquad\qquad \xi \longmapsto \mathbb{C}\begin{bmatrix}1\\\xi\end{bmatrix} \subset \mathbb{C}^2$$

where $\zeta\xi = 1$ on the intersection $\mathbb{C}^* \cong \mathbb{A}_1 \cap \mathbb{A}_2$.

Stereographic projection. The unit sphere S^2 of purely imaginary unit quaternions, with its complex structure defined in §5.1, identifies with $\mathbb{P}^1$ as follows. We embed $\mathbb{C} \subset \mathbb{H}$ as the subalgebra generated by I. The two affine patches $\mathbb{A}_1, \mathbb{A}_2 \subset \mathbb{P}^1$ map affinely into the multiplicative subgroup $\mathbb{H}^*$ of nonzero quaternions by:

$$\mathbb{A}_1 \longrightarrow \mathbb{H}^*$$
$$\psi^{(1)}(\zeta) \longmapsto 1 + \zeta K$$

and

$$\mathbb{A}_2 \longrightarrow \mathbb{H}^*$$
$$\psi^{(2)}(\xi) \longmapsto \bar{\xi} + K$$

respectively. Use these maps to conjugate $I \in \mathbb{H}$ to purely imaginary unit quaternions in S^2:

$$I^{(1)}_\zeta := (1 + \zeta K)I(1 + \zeta K)^{-1}$$
(6.1.1) $$I^{(2)}_\xi := (\bar{\xi} + K)I(\bar{\xi} + K)^{-1} = (\xi - K)^{-1}I(\xi - K),$$

where $\zeta, \xi \in \mathbb{C}$. These two maps combine to define a map $\mathbb{P}^1 \longrightarrow S^2$ because:

LEMMA 6.1.1. *If $\xi\zeta = 1$, then $I^{(1)}_\zeta = I^{(2)}_\xi$.*

PROOF. Since $\zeta \in \mathbb{C} = \mathbb{R}1 + \mathbb{R}I$,

$$1 + \zeta K = 1 + K\bar{\zeta} = (\bar{\zeta}^{-1} + K)\bar{\zeta} = (\bar{\xi} + K)\bar{\zeta}$$

so

$$(1 + \zeta K)^{-1} = (\bar{\zeta})^{-1}(\bar{\xi} + K)^{-1}$$

and

$$\begin{aligned}I^{(1)}_\zeta &= (1 + \zeta K)I(1 + \zeta K)^{-1}\\&= (\bar{\xi} + K)\bar{\zeta}I(\bar{\zeta})^{-1}(\bar{\xi} + K)^{-1}\\&= (\bar{\xi} + K)I(\bar{\xi} + K)^{-1}\\&= I^{(2)}_\xi.\end{aligned}$$

□

6. THE TWISTOR SPACE

Evidently $I_0^{(1)} = I$, $I_0^{(2)} = -I$, and $I^{(1)}(1) = I^{(2)}(1) = J$. Explicitly, $I_\zeta^{(1)}, I_\xi^{(2)}$ are *stereographic projections*:

(6.1.2)
$$I_\zeta^{(1)} = \frac{1 - \zeta\bar\zeta}{1 + \zeta\bar\zeta}I + \frac{2\operatorname{Re}(\zeta)}{1 + \zeta\bar\zeta}J + \frac{2\operatorname{Im}(\zeta)}{1 + \zeta\bar\zeta}K$$
$$I_\xi^{(2)} = -\frac{1 - \xi\bar\xi}{1 + \xi\bar\xi}I + \frac{2\operatorname{Re}(\xi)}{1 + \xi\bar\xi}J - \frac{2\operatorname{Im}(\xi)}{1 + \xi\bar\xi}K.$$

(This differs slightly from formulas (3.70) and (3.71) of [**HKLR**]. The formulas in [**HKLR**] relate to ours by replacing ζ with $\bar\zeta$. In our convention, stereographic projection is a map from $\mathbb{P}^1$ to S^2 with the orientation of S^2 given by the *outward* normal vector.)

The antipodal map. Stereographic projection relates the antipodal map $S^2 \longrightarrow S^2$ defined in (5.1.5) to the map $\mathbb{P}^1 \xrightarrow{\alpha} \mathbb{P}^1$ induced by the anti-linear involution of $\mathbb{C}^2$:

$$\mathbb{C}^2 \xrightarrow{\bar\alpha} \mathbb{C}^2$$
$$\begin{bmatrix} z_1 \\ z_2 \end{bmatrix} \longmapsto \begin{bmatrix} -\bar z_2 \\ \bar z_1 \end{bmatrix}.$$

In terms of the atlas, it is given by:

$$\psi^{(1)}(\zeta) \xleftrightarrow{\alpha} \psi^{(2)}(-\bar\zeta)$$
$$\psi^{(2)}(\xi) \xleftrightarrow{\alpha} \psi^{(1)}(-\bar\xi)$$

for $\xi, \zeta \in \mathbb{C}$ and in particular

$$\psi^{(1)}(0) \xleftrightarrow{\alpha} \psi^{(2)}(0).$$

LEMMA 6.1.2. *The antipodal map takes a complex structure to its opposite. That is, if $\zeta \in \mathbb{C}$ is nonzero, then*

$$I_{-\bar\zeta^{-1}}^{(1)} = I_{-\bar\zeta}^{(2)} = -I_\zeta^{(1)}$$

PROOF.

$$\begin{aligned}
I^{(1)}_{-\bar\zeta^{-1}} &= (1-\bar\zeta^{-1}K)I(1-\bar\zeta^{-1}K)^{-1}\\
&= \frac{1}{1+|\bar\zeta^{-1}|^2}(1-\bar\zeta^{-1}K)I(1+\bar\zeta^{-1}K)\\
&= \frac{|\zeta|^2}{|\zeta|^2+1}(1-\bar\zeta^{-1}K)I(1+\bar\zeta^{-1}K)\\
&= \frac{1}{1+|\zeta|^2}(1-K\zeta^{-1})\zeta I\bar\zeta(1+\bar\zeta^{-1}K)\\
&= \frac{1}{1+|\zeta|^2}(\zeta-K)I(\zeta+K)\\
&= \frac{1}{1+|\zeta|^2}(-\zeta K-1)KIK(1-\zeta K)\\
&= -\frac{1}{1+|\zeta|^2}(1+\zeta K)I(1-\zeta K)\\
&= -I^{(1)}_\zeta
\end{aligned}$$

as claimed. $\square$

Line bundles. For each $d \in \mathbb{Z}$ there is a holomorphic line bundle $\mathcal{O}(d)$ of degree d over $\mathbb{P}^1$, unique up to isomorphism. Every holomorphic line bundle over $\mathbb{P}^1$ is isomorphic to some $\mathcal{O}(d)$.

Explicitly, the total space of $\mathcal{O}(d)$ is defined in the above atlas as the identification space of the disjoint union

$$\left(\mathbb{C}\times \mathbb{A}_1\right)\coprod\left(\mathbb{C}\times \mathbb{A}_2\right)$$

under the equivalence relation:

(6.1.3) $$\left(z,\psi^{(1)}(\zeta),z\right) \sim \left(\xi^d z, \psi^{(2)}(\xi)\right),$$

where $\zeta\xi = 1$. A section s of $\mathcal{O}(d)$ determines a pair $\left(s^{(1)}, s^{(2)}\right)$ of maps $\mathbb{C} \longrightarrow \mathbb{C}$ such that $s\bigl(\psi^{(1)}(\zeta)\bigr)$ corresponds to $\bigl(\psi^{(1)}(\zeta), s^{(1)}(\zeta)\bigr)$ in $\mathbb{A}_1 \times \mathbb{C}$ and $s\bigl(\psi^{(2)}(\xi)\bigr)$ corresponds to $\bigl(\psi^{(2)}(\xi), s^{(2)}(\xi)\bigr)$ in $\mathbb{A}_2 \times \mathbb{C}$. By (6.1.3), the pair $\left(s^{(1)}, s^{(2)}\right)$ must satisfy:

(6.1.4) $$s^{(1)}(\zeta) = \xi^{-d} s^{(2)}(\xi)$$

whenever $\zeta\xi = 1$. Conversely any pair $\left(s^{(1)}, s^{(2)}\right)$ satisfying (6.1.4) corresponds to a section of $\mathcal{O}(d)$

The trivial rank two bundle

$$\mathbb{C}^2 \times \mathbb{P}^1 \longrightarrow \mathbb{P}^1$$

contains the (tautological) subbundle $\mathcal{O}(-1)$ whose fiber over $[v] \in \mathbb{P}^1$ is the line $\mathbb{C}v \subset \mathbb{C}^2$. The *hyperplane bundle* over $\mathbb{P}^1$ is its dual $\mathcal{O}(1)$: the fiber over $\mathbb{C}v$ consists of linear functionals on $\mathbb{C}_v$. A linear functional

$$\mathbb{C}^2 \xrightarrow{a} \mathbb{C}$$

defines a section of $\mathcal{O}(1)$ by assigning to $[v] \in \mathbb{P}^1$ its restriction to $\mathbb{C}v$. Every holomorphic section of $\mathcal{O}(1)$ arises from such a linear functional on $\mathbb{C}^2$. A nonzero holomorphic section has a single zero of order 1, corresponding to the kernel of a. For example, the linear functional

$$\begin{bmatrix} z_1 \\ z_2 \end{bmatrix} \longmapsto z_2$$

determines the section σ of $\mathcal{O}(1)$ having

$$s^{(1)}(\zeta) = 1, \quad s^{(2)}(\xi) = \xi.$$

This section is holomorphic on $\mathbb{A}_1$ and has a single simple zero at $\psi^{(2)}(0) \in \mathbb{A}_2$.

The *canonical line bundle* $T^*\mathbb{P}^1$ has degree -2 and is isomorphic to $\mathcal{O}(-2)$; the holomorphic 1-form

$$d\zeta = -\xi^{-2} d\xi$$

is holomorphic on $\mathbb{A}_1^{\cdot}$ with a double pole at $\zeta = \infty$. In the atlas, this section is given by holomorphic functions

$$s^{(1)}(\zeta) = 1$$
$$s^{(2)}(\xi) = \xi^{-2}$$

where the local nonvanishing sections over $\mathbb{A}_1$ and $\mathbb{A}_2$ are $d\zeta$ and $-d\xi$ respectively.

The inverse of the canonical bundle, the *tangent bundle* $T\mathbb{P}^1 \cong \mathcal{O}(2)$, has a holomorphic section

(6.1.5) $$\delta = \frac{\partial}{\partial \zeta} = -\xi^2 \frac{\partial}{\partial \xi}$$

with a double zero at $\xi = 0$. In coordinate charts,

$$\delta^{(1)}(\zeta) = 1$$
$$\delta^{(2)}(\xi) = \xi^2.$$

This section corresponds to the section $-\sigma^2 = -\sigma \otimes \sigma$ of $\mathcal{O}(2) = \mathcal{O}(1) \otimes \mathcal{O}(1)$.

Antiholomorphic line bundles. Given any holomorphic vector bundle $E \longrightarrow M$, the *conjugate* vector bundle $\bar{E} \longrightarrow M$ is defined as the holomorphic vector bundle with opposite complex structure on the fibers. Fiberwise complex-conjugation defines an isomorphism of smooth vector bundles between E and $\bar{E}$ which is *antiholomorphic*. A Hermitian structure on $\mathbb{C}^2$ defines Hermitian structures on any $\mathcal{O}(d)$ which induces isomorphisms

$$\overline{\mathcal{O}(d)} \cong \mathcal{O}(-d).$$

LEMMA 6.1.3. *Suppose that $d \in \mathbb{Z}$ is even. Then the antipodal map $\boldsymbol{\alpha}$ of $\mathbb{P}^1$ lifts to an antiholomorphic isomorphism*

$$\mathcal{O}(d) \xrightarrow{\widetilde{\boldsymbol{\alpha}}} \overline{\mathcal{O}(d)}.$$

PROOF. In the coordinate atlas, define

$$\left(z, \psi^{(1)}(\zeta)\right) \xmapsto{\widetilde{\boldsymbol{\alpha}}} \left(\bar{z}, \psi^{(2)}(-\bar{\zeta})\right)$$
$$\left(w, \psi^{(2)}(\xi)\right) \xmapsto{\widetilde{\boldsymbol{\alpha}}} \left(\bar{w}, \psi^{(1)}(-\bar{\xi})\right).$$

Suppose that $z, \zeta \in \mathbb{C}$ define a point $\left(z, \psi^{(1)}(\zeta)\right)$ in $\mathbb{C} \times \mathbb{A}_1$, whose image $\widetilde{\boldsymbol{\alpha}}\left(z, \psi^{(1)}(\zeta)\right)$ corresponds to

(6.1.6) $$\left((-\bar{\xi})^d \bar{z},\ \psi^{(1)}(-\bar{\xi})\right) \in \mathbb{C} \times \mathbb{A}_1,$$

where $\zeta \xi = 1$. On the other hand, $\left(z, \psi^{(1)}(\zeta)\right)$ corresponds to the point $\left(\xi^d z, \psi^{(2)}(\xi)\right)$ in $\mathbb{C} \times \mathbb{A}_2$. This point maps under $\widetilde{\boldsymbol{\alpha}}$ to $\left(\overline{(\xi^d z)}, \psi^{(1)}(-\bar{\xi})\right)$, which equals (6.1.6) when d is even. $\square$

When $d = 2$, the image $(\widetilde{\boldsymbol{\alpha}})^* d\zeta = -\overline{d\xi}$ is the image of the section defined above under the differential of $\boldsymbol{\alpha}$.

6.2. The twistor space as a smooth vector bundle.

The triviality of the vector bundle $\mathbb{C}^{2k} \otimes \mathcal{O}(1)$ as a *smooth real vector bundle* may be seen as follows. Isomorphism classes of bundles over $\mathbb{P}^1$ with structure group G are classified by the homotopy class of their *clutching function* $\mathbb{A}_1 \cap \mathbb{A}_2 \longrightarrow G$, which identifies with an element of $\pi_1(G)$ (see for example Steenrod [**S**]). Namely, a fiber bundle over a contractible space is trivial, and writing a trivialization over $\mathbb{A}_1$ in terms of a trivialization over $\mathbb{A}_2$ defines a map $\mathbb{A}_1 \cap \mathbb{A}_2 \longrightarrow G$, whose homotopy class is a complete invariant of the isomorphism type as a smooth fiber bundle with structure group G. Since $\mathbb{A}_1 \cap \mathbb{A}_2$ is homotopy-equivalent to S^1, isomorphism classes of G-bundles are classified by $\pi_1(G)$.

The invariant of the line bundle $\mathcal{O}(1)$ is a generator $\alpha \in \pi_1(\mathrm{U}(1))$. The invariant of $\mathbb{C}^{2k} \otimes \mathcal{O}(1)$ is the image of
$$\alpha \oplus \cdots \oplus \alpha \in \pi_1\big(\mathrm{U}(1) \times \cdots \times \mathrm{U}(1)\big)$$
under the homomorphism
$$\begin{array}{ccc} \pi_1\big(\mathrm{U}(1) \times \cdots \times \mathrm{U}(1)\big) & \longrightarrow & \pi_1\big(\mathrm{U}(2k)\big) \\ \| & & \| \\ \mathbb{Z}^{2k} & \longrightarrow & \mathbb{Z} \end{array}$$
induced by the inclusion of diagonal matrices $\mathrm{U}(1) \times \cdots \times \mathrm{U}(1)$ in $\mathrm{U}(2k)$. The isomorphism $\pi_1(\mathrm{U}(2k)) \longrightarrow \mathbb{Z}$ is induced by
$$\det : \mathrm{U}(2k) \longrightarrow \mathrm{U}(1).$$
Thus the invariant of $\mathbb{C}^{2k} \otimes \mathcal{O}(1)$ as a complex vector bundle equals $2k\alpha$, where α denotes a generator of $\pi_1\big(\mathrm{U}(2k)\big) \cong \mathbb{Z}$. The isomorphism type *as a real vector bundle* is the image of $2k\alpha$ under the homomorphism
$$\begin{array}{ccc} \pi_1(\mathrm{U}(2k)) & \longrightarrow & \pi_1(\mathrm{O}(4k)) \\ \| & & \| \\ \mathbb{Z} & \longrightarrow & \mathbb{Z}/2 \end{array}$$
which is evidently zero.

6.3. A holomorphic atlas for the twistor space. The twistor space of the quaternionic vector space $V_{\mathbb{H}}$ is defined as the product $\tilde{Z} = V_{\mathbb{H}} \times \mathbb{P}^1$ with coordinate projection:
$$\tilde{Z} = V_{\mathbb{H}} \times \mathbb{P}^1 \xrightarrow{\Pi_{\tilde{Z}}} \mathbb{P}^1.$$
The complex structure on $\tilde{Z}$ is the unique complex structure such that $\Pi_{\tilde{Z}}$ is a holomorphic vector bundle, and the complex structure on the fiber over $[v] \in \mathbb{P}^1$ is the complex structure defined by left-multiplication by the purely imaginary unit quaternion corresponding to $[v]$. We shall see that the twistor space $\tilde{Z}$ of $V_{\mathbb{H}}$ is the total space of the holomorphic vector bundle $\mathbb{C}^{2k} \otimes \mathcal{O}(1)$, and describe the structures (holomorphic symplectic structures, real structure) on this space explicitly in terms of the atlas $\{(\mathbb{A}_1, \psi^{(1)}), (\mathbb{A}_2, \psi^{(2)})\}$ in this and subsequent sections.

Represent complex vectors $\mathbf{v} \in \mathbb{C}^{2k}$ by the notation:

(6.3.1)
$$\mathbf{v} = \begin{bmatrix} \mathsf{q} \\ \mathsf{p} \end{bmatrix}$$

where the components are vectors $\mathsf{q}, \mathsf{p} \in \mathbb{C}^k$. Thus a general quaternion vector in $\mathbb{H}^k$ is:

$$\mathsf{q} + \mathsf{p}J = \mathsf{x}_1 + \mathsf{y}_1 I + \mathsf{x}_2 J + \mathsf{y}_2 K,$$

where $\mathsf{q} = \mathsf{x}_1 + i\mathsf{y}_1$ and $\mathsf{p} = \mathsf{x}_2 + i\mathsf{y}_2$ for real vectors $\mathsf{x}_1, \mathsf{x}_2, \mathsf{y}_1, \mathsf{y}_2 \in \mathbb{R}^k$.

For $\zeta, \xi \in \mathbb{C}, \mathsf{v} \in \mathbb{C}^{2k}$, define

$$\mathbb{C}^{2k} \xrightarrow{f_\zeta^{(1)}} \mathbb{H}^k$$

$$\mathsf{v} = \begin{bmatrix} \mathsf{q} \\ \mathsf{p} \end{bmatrix} \longmapsto \frac{1}{1+\zeta\bar{\zeta}}(1+\zeta K)(\mathsf{q}+\mathsf{p}J) = (1-\zeta K)^{-1}(\mathsf{q}+\mathsf{p}J)$$

and

$$\mathbb{C}^{2k} \xrightarrow{f_\xi^{(2)}} \mathbb{H}^k$$

$$\mathsf{v} = \begin{bmatrix} \mathsf{q} \\ \mathsf{p} \end{bmatrix} \longmapsto \frac{1}{1+\xi\bar{\xi}}(\bar{\xi}+K)(\mathsf{q}+\mathsf{p}J) = (\xi-K)^{-1}(\mathsf{q}+\mathsf{p}J).$$

For each $\xi, \zeta \in \mathbb{C}$, the maps $f_\zeta^{(1)}, f_\xi^{(2)}$, are $\mathbb{R}$-linear isomorphisms.

Hence if we define charts

$$\mathbb{C}^{2k} \times \mathbb{A}_1 \xrightarrow{F^{(1)}} \mathbb{H}^k \times \mathbb{A}_1$$

$$\mathbb{C}^{2k} \times \mathbb{A}_2 \xrightarrow{F^{(2)}} \mathbb{H}^k \times \mathbb{A}_2$$

(6.3.2)
$$F^{(1)}(\mathsf{v},\zeta) = \left(f_\zeta^{(1)}(\mathsf{v}), \psi^{(1)}(\zeta)\right)$$
$$F^{(2)}(\mathsf{v},\xi) = \left(f_\xi^{(2)}(\mathsf{v}), \psi^{(2)}(\xi)\right),$$

we have

$$F^{(1)}\bigl(\mathsf{v}, \psi^{(1)}(\zeta)\bigr) = F^{(2)}\bigl(\xi\mathsf{v}, \psi^{(2)}(\xi)\bigr),$$

thus, obtaining a holomorphic rank $2k$ vector bundle

$$\Pi_{\tilde{Z}} : \tilde{Z} \longrightarrow \mathbb{P}^1.$$

The fiber of $\tilde{Z}_\zeta$ over $\psi^{(1)}(\zeta)$ is $\mathbb{H}^k$ with the complex structure $I_\zeta^{(1)}$. We denote the fiber $\Pi_{\tilde{Z}}^{-1}\bigl(\psi^{(2)}(0)\bigr)$ by $\tilde{Z}_\infty$.

LEMMA 6.3.1. *Let* $\xi\zeta = 1$. *Then* $f_\zeta^{(1)}(\mathsf{v}) = f_\xi^{(2)}(\xi\mathsf{v})$.

PROOF.

$$\begin{aligned}
f^{(1)}_\zeta(\mathsf{v}) &= \frac{1}{1+\zeta\bar\zeta}(1+\zeta K)(\mathsf{q}+\mathsf{p}J)\\
&= \frac{\xi\bar\xi}{\xi\bar\xi+1}(1+\xi^{-1}K)(\mathsf{q}+\mathsf{p}J)\\
&= \frac{1}{1+\xi\bar\xi}(1+\xi^{-1}K)\xi\bar\xi(\mathsf{q}+\mathsf{p}J)\\
&= \frac{1}{\xi\bar\xi+1}(1+K\bar\xi^{-1})\bar\xi\,\xi(\mathsf{q}+\mathsf{p}J)\\
&= \frac{1}{\xi\bar\xi+1}(\bar\xi+K)\xi(\mathsf{q}+\mathsf{p}J)\\
&= f^{(2)}_\xi(\xi\mathsf{v})
\end{aligned}$$

□

Thus $\{F^{(1)}, F^{(2)}\}$ is an atlas for a holomorphic vector bundle $\tilde Z$ of rank $2k$ over $\mathbb{P}^1$. Lemma 6.3.1 implies that this vector bundle is $\mathbb{C}^{2k}\otimes\mathcal{O}(1)$, a direct sum of $2k$ copies of $\mathcal{O}(1)$.

Finally, the complex structure on the fiber $\tilde Z_\zeta$ is $I^{(1)}_\zeta$ and $I^{(2)}_0$ if $\zeta=\infty$.

LEMMA 6.3.2. *Let* $\zeta,\xi\in\mathbb{C}$, $\zeta\xi=1$. *Then*
$$f^{(1)}_\zeta(i\mathsf{v}) = I^{(1)}_\zeta f^{(1)}_\zeta(\mathsf{v}), \qquad f^{(2)}_\xi(i\mathsf{v}) = I^{(2)}_\xi f^{(2)}_\xi(\mathsf{v}).$$

PROOF.

$$\begin{aligned}
f^{(1)}_\zeta(i\mathsf{v}) &= (1+\zeta\bar\zeta)^{-1}(1+\zeta K)(i\mathsf{q}+i\mathsf{p}J)\\
&= (1+\zeta\bar\zeta)^{-1}(1+\zeta K)I(\mathsf{q}+\mathsf{p}J)\\
&= (1+\zeta\bar\zeta)^{-1}I^{(1)}_\zeta(1+\zeta K)(\mathsf{q}+\mathsf{p}J)\\
&= I^{(1)}_\zeta f^{(1)}_\zeta(\mathsf{v})
\end{aligned}$$

and similarly for $f^{(2)}_\xi$.

□

6.4. The twistor lines. By (6.3.2) and Lemma 6.3.1, each $\mathsf{v}\in\mathbb{C}^{2k}$ defines a nowhere vanishing holomorphic section of l_v, called the *twistor line* corresponding to v. The normal bundle to l_v is isomorphic to the vector bundle $\Pi_{\tilde Z}$ itself, that is, to $\mathbb{C}^{2k}\otimes\mathcal{O}(1)$.

Here is how the the twistor lines appear in the holomorphic atlas. Write
$$\mathsf{v}_0 = \mathsf{q}_0 + \mathsf{p}_0 J$$

with $\mathsf{q}_0, \mathsf{p}_0 \in \mathbb{C}^k$. The twistor line $l_v(\psi^{(1)}(\zeta))$ intersects the fiber $\tilde{Z}_\zeta$ at
$$F^{(1)}\left(\begin{bmatrix} \mathsf{q}^{(1)}(\zeta) \\ \mathsf{p}^{(1)}(\zeta) \end{bmatrix}, \psi^{(1)}(\zeta)\right),$$
where $\mathsf{q}^{(1)}(\zeta), \mathsf{p}^{(1)}(\zeta) \in \mathbb{C}^k$ are defined by
(6.4.1) $$\begin{bmatrix} \mathsf{q}^{(1)}(\zeta) \\ \mathsf{p}^{(1)}(\zeta) \end{bmatrix} := \begin{bmatrix} \mathsf{q}_0 + i\zeta \bar{\mathsf{p}}_0 \\ \mathsf{p}_0 - i\zeta \bar{\mathsf{q}}_0 \end{bmatrix}.$$

Similarly, the twistor line $l_v(\psi^{(2)}(\xi))$ intersects $\tilde{Z}_{\xi^{-1}}$ at
$$F^{(2)}\left(\begin{bmatrix} \mathsf{q}^{(2)}(\xi) \\ \mathsf{p}^{(2)}(\xi) \end{bmatrix}, \psi^{(2)}(\xi)\right)$$
where
(6.4.2) $$\begin{bmatrix} \mathsf{q}^{(2)}(\xi) \\ \mathsf{p}^{(2)}(\xi) \end{bmatrix} := \begin{bmatrix} \xi \mathsf{q}_0 + i\bar{\mathsf{p}}_0 \\ \xi \mathsf{p}_0 - i\bar{\mathsf{q}}_0 \end{bmatrix}.$$

Conversely, the twistor line containing
$$F^{(1)}\left(\begin{bmatrix} \mathsf{q} \\ \mathsf{p} \end{bmatrix}, \psi^{(1)}(\zeta)\right) \in \Pi_{\tilde{Z}}^{-1}(\mathbb{A}_1)$$
intersects $\tilde{Z}_0$ at
$$F^{(1)}\left(\frac{1}{1+\zeta\bar{\zeta}} \begin{bmatrix} \mathsf{p} + i\zeta\bar{\mathsf{q}} \\ \mathsf{q} - i\zeta\bar{\mathsf{p}} \end{bmatrix}, \psi^{(1)}(0)\right)$$
and the twistor line containing
$$F^{(2)}\left(\begin{bmatrix} \mathsf{q} \\ \mathsf{p} \end{bmatrix}, \psi^{(2)}(\xi)\right) \in \Pi_{\tilde{Z}}^{-1}(\mathbb{A}_2)$$
intersects $\tilde{Z}_\infty$ at
$$F^{(2)}\left(\frac{1}{1+\xi\bar{\xi}} \begin{bmatrix} \bar{\xi}\mathsf{p} + i\bar{\mathsf{q}} \\ \bar{\xi}\mathsf{q} - i\bar{\mathsf{p}} \end{bmatrix}, \psi^{(2)}(0)\right).$$

6.5. The real structure on the twistor space. The antipodal map $\mathbb{P}^1 \xrightarrow{\alpha} \mathbb{P}^1$ lifts to the twistor space $\tilde{Z}$ preserving the fibration and the twistor lines. In terms of the product fibration in the coordinate charts, the lift $\tilde{\alpha}$ to $\tilde{Z}$ is given by:

$$F^{(1)}\left(\begin{bmatrix} \mathsf{q} \\ \mathsf{p} \end{bmatrix}, \psi^{(1)}(\zeta)\right) \xleftrightarrow{\tilde{\alpha}} F^{(2)}\left(\begin{bmatrix} -i\bar{\xi}\bar{\mathsf{p}} \\ i\bar{\xi}\bar{\mathsf{q}} \end{bmatrix}, \psi^{(2)}(-\bar{\xi})\right)$$

where $\xi\zeta = 1$ and

(6.5.1) $$F^{(1)}\left(\begin{bmatrix} \mathsf{q} \\ \mathsf{p} \end{bmatrix}, \psi^{(1)}(0)\right) \xleftrightarrow{\tilde{\alpha}} F^{(2)}\left(\begin{bmatrix} -i\bar{\mathsf{p}} \\ i\bar{\mathsf{q}} \end{bmatrix}, \psi^{(2)}(0)\right).$$

Lemma 6.1.2 and the antiholomorphicity of $\boldsymbol{\alpha}$ imply that $\widetilde{\boldsymbol{\alpha}}$ is an antiholomorphic involution of Z.

6.6. Symplectic geometry of the twistor space. The complex-symplectic structures defined in (5.2.3) fit together giving a holomorphic exterior 2-form on the complex manifold $\boldsymbol{\Omega}$ on Z, taking values in the holomorphic line bundle $\Pi_{\tilde{Z}}^{*}\mathcal{O}(2)$, where $\Pi_{\tilde{Z}}$ denotes the twistor fibration.

Let $u \in S^2$. There is a family of complex-symplectic structures on $V_{\mathbb{H}}$ which are holomorphic with respect to u, which are parametrized by a unit vector $u_1 \in T_1 S^2$. Namely (5.2.3) defines a complex-symplectic structure $\Omega_{(u_1, u_2)}$, where $u_1 \in u^{\perp}$ and $u_2 = u \times u_1$. Let u be $I_\zeta^{(1)}$ or $I_\xi^{(2)}$. Define u_1 (respectively u_2) as $J_\zeta^{(1)}, J_\xi^{(2)}$ (respectively $K_\zeta^{(1)}, K_\xi^{(2)}$), where:

$$J_\zeta^{(1)} := (1 + \zeta K)J(1 + \zeta K)^{-1}$$
$$K_\zeta^{(1)} := (1 + \zeta K)K(1 + \zeta K)^{-1}$$
$$J_\xi^{(2)} := (\bar{\xi} + K)J(\bar{\xi} + K)^{-1}$$
(6.6.1) $$K_\xi^{(2)} := (\bar{\xi} + K)K(\bar{\xi} + K)^{-1},$$

analogous to the definition (6.1.1) of $I_\zeta^{(1)}$ and $I_\xi^{(2)}$.

Evidently, $(J_0^{(1)}, K_0^{(1)}) = (J, K)$ and $(J_0^{(2)}, K_0^{(2)}) = (-J, K)$. For $\zeta, \xi \in \mathbb{C}$, let

$$\boldsymbol{\Omega}_\zeta^{(1)} := (1 + \zeta\bar{\zeta})\Omega_{(J_\zeta^{(1)}, K_\zeta^{(1)})}$$
$$\boldsymbol{\Omega}_\xi^{(2)} := (1 + \xi\bar{\xi})\Omega_{(J_\xi^{(2)}, K_\xi^{(2)})}.$$

Corresponding to $\psi^{(1)}(0) \in \mathbb{A}_1$ and $\psi^{(2)}(0) \in \mathbb{A}_2$ are

$$\boldsymbol{\Omega}_0^{(1)} = \Omega_{(J,K)}, \quad \boldsymbol{\Omega}_0^{(2)} = \Omega_{(-J,K)},$$

respectively.

LEMMA 6.6.1. *Let α and β be tangent vectors and $\zeta, \xi \in \mathbb{C}$. Then*

$$\boldsymbol{\Omega}_\zeta^{(1)}(\alpha, \beta) = \Omega_{(J,K)}\big((1 - \zeta K)\alpha, (1 - \zeta K)\beta\big)$$
$$\boldsymbol{\Omega}_\xi^{(2)}(\alpha, \beta) = \Omega_{(J,K)}\big((\xi - K)\alpha, (\xi - K)\beta\big)$$

PROOF. First recall the metric g as defined in (5.2.1). For any α and β,

$$g\big((1 + \zeta K)\alpha, \beta\big) = g\big(\alpha, (1 - \zeta K)\beta\big)$$
$$g\big((\bar{\xi} + K)\alpha, \beta\big) = g\big(\alpha, (\xi - K)\beta\big)$$

since $g(h\alpha, \beta) = g(\alpha, -h\beta)$ for any purely imaginary quaternion $h \in \mathbb{H}$.

$$\begin{aligned}
\mathbf{\Omega}^{(1)}_\zeta(\alpha,\beta) &= (1+\zeta\bar\zeta)\left(g(J^{(1)}_\zeta \alpha, \beta) + ig(K^{(1)}_\zeta \alpha, \beta)\right) \\
&= (1+\zeta\bar\zeta)\left(g\big((1+\zeta K)J(1+\zeta K)^{-1}\alpha, \beta\big) \right.\\
&\qquad\qquad\left. + ig\big((1+\zeta K)K(1+\zeta K)^{-1}\alpha, \beta\big)\right) \\
&= g\big((1+\zeta K)J(1-\zeta K)\alpha, \beta\big) + ig\big((1+\zeta K)K(1-\zeta K)\alpha, \beta\big) \\
&= g\big(J(1-\zeta K)\alpha, (1-\zeta K)\beta\big) + ig\big(K(1-\zeta K)\alpha, (1-\zeta K)\beta\big) \\
&= \Omega_{(J,K)}\big((1-\zeta K)\alpha, (1-\zeta K)\beta\big)
\end{aligned}$$

and similarly for $\mathbf{\Omega}^{(2)}_\xi(\alpha,\beta)$. □

COROLLARY 6.6.2. *If $\zeta\xi = 1$, then $\mathbf{\Omega}^{(2)}_\xi = \xi^2 \mathbf{\Omega}^{(1)}_\zeta$.*

PROOF. Since $\xi - K = \xi(1-\zeta K)$ and $\Omega_{(J,K)}$ is holomorphic with respect to I,

$$\begin{aligned}
\mathbf{\Omega}^{(2)}_\xi(\alpha,\beta) &= \Omega_{(J,K)}\big((\xi - K)\alpha, (\xi - K)\beta\big) \\
&= \xi^2 \Omega_{(J,K)}\big((1-\zeta K)\alpha, (1-\zeta K)\beta\big) \\
&= \xi^2 \mathbf{\Omega}^{(1)}_\zeta(\alpha,\beta)
\end{aligned}$$

as desired. □

Thus $\{\mathbf{\Omega}^{(1)}, \mathbf{\Omega}^{(2)}\}$ defines a $\Pi^*_{\tilde Z}\mathcal{O}(2)$-valued exterior 2-form $\mathbf{\Omega}$ on $\tilde Z$. Denoting δ the meromorphic section of $\mathcal{O}(-2)$ in (6.1.5), we see that $\mathbf{\Omega} \otimes \Pi^*_{\tilde Z}\delta$ is a (scalar-valued) exterior 2-form on $\tilde Z$.

COROLLARY 6.6.3. *$\mathbf{\Omega}$ is holomorphic.*

PROOF. $\mathbf{\Omega}$ is holomorphic in the fiber directions since $\mathbf{\Omega}^{(1)}_\zeta(I^{(1)}_\zeta(\alpha), \beta) = i\mathbf{\Omega}^{(1)}_\zeta(\alpha,\beta)$ on $\mathbb{H}^k \times \mathbb{A}_1$. Now

$$\begin{aligned}
\mathbf{\Omega}^{(1)}_\zeta(\alpha,\beta) &= \Omega_{(J,K)}\big((1-\zeta K)\alpha, (1-\zeta K)\beta\big) \\
&= \Omega_{(J,K)}(\alpha,\beta) \\
&\quad + \zeta\left(\Omega_{(J,K)}(K\alpha, \beta) + \Omega_{(J,K)}(\alpha, K\beta)\right) \\
&\quad + \zeta^2 \Omega_{(J,K)}(K\alpha, K\beta) \\
&= \Omega_{(J,K)}(\alpha,\beta) - 2\zeta\,\omega_I(\alpha,\beta) + \zeta^2 \Omega_{(-J,K)}(\alpha,\beta).
\end{aligned}$$

That is,

(6.6.2) $$\mathbf{\Omega}^{(1)}_\zeta = \Omega_{(J,K)} - 2\zeta\omega_I + \zeta^2 \Omega_{(-J,K)};$$

therefore, $\mathbf{\Omega}^{(1)}$ is holomorphic with respect to ζ.

Similarly, $\mathbf{\Omega}^{(2)}_\xi(I^{(2)}_\xi(\alpha),\beta) = i\mathbf{\Omega}^{(2)}_\xi(\alpha,\beta)$ on $\mathbb{H}^k \times \mathbb{A}_2$, and

(6.6.3) $$\mathbf{\Omega}^{(2)}_\xi = \Omega_{(-J,K)} - 2\xi\omega_I + \xi^2 \Omega_{(J,K)}.$$

$\square$

(Formulas (6.6.2),(6.6.3) differ slightly from formula (3.87) of [**HKLR**], due to our differing conventions; see the discussion after (6.1.2).)

LEMMA 6.6.4. *Let $\tilde{\boldsymbol{\alpha}}$ be the real structure on $\tilde{Z}$ and $\mathbf{\Omega}$ the $\mathcal{O}(2)$-valued holomorphic exterior 2-form defined above. Then*

$$(\tilde{\boldsymbol{\alpha}})^*\mathbf{\Omega} = -\overline{\mathbf{\Omega}}.$$

PROOF. The action of $\tilde{\boldsymbol{\alpha}}$ on the coefficient bundle $\mathcal{O}(2)$ is defined in Lemma 6.1.3. Now

$$\mathbf{\Omega}^{(1)}_0 = \Omega_{(J,K)} = -\overline{\Omega_{(-J,K)}} = -\overline{\mathbf{\Omega}^{(2)}_0}.$$

Let $\xi\zeta = 1$. Applying (6.6.2),

$$\mathbf{\Omega}^{(1)}_{-\bar{\xi}} = -\bar{\xi}^2 \mathbf{\Omega}^{(1)}_\zeta.$$

$\square$

6.7. The lattice quotient. We have described the essential structures on the twistor space $\tilde{Z} \longrightarrow \mathbb{P}^1$ of the quaternionic vector space $V_\mathbb{H}$. That is $\tilde{Z} \cong \mathbb{H}^k \times \mathbb{P}^1$ with the complex structure which restricts to the fiber over $[v] \in \mathbb{P}^1$ by the corresponding complex structure defined by (6.1.2). The coordinate charts $F^{(i)}$ defined in (6.3.2) trivialize the holomorphic vector bundle $\tilde{Z}$ over the affine patches $\mathbb{A}_1, \mathbb{A}_2$ of $\mathbb{P}^1$.

The twistor space for the hyperkähler manifold $V_\mathbb{H}/L$ is the quotient $Z = \tilde{Z}/L$ with the fibration $Z \xrightarrow{\Pi_Z} \mathbb{P}^1$ induced from $\Pi_{\tilde{Z}}$. Denote the fiber $\Pi_Z^{-1}(\psi^{(1)}(\zeta))$ by Z_ζ and the fiber $\Pi_Z^{-1}(\psi^{(2)}(0))$ by Z_∞, similar to $\tilde{Z}_\zeta$ and $\tilde{Z}_\infty$. The complex structure on $\tilde{Z}$, the complex $\Pi_{\tilde{Z}}^*\mathcal{O}(2)$-valued exterior form $\mathbf{\Omega}$ and the real structure $\tilde{\boldsymbol{\alpha}}$ are all invariant under translations in L, and therefore induce corresponding structures on the quotient $Z = \tilde{Z}/L$.

The collection of twistor lines l_v is invariant under the action of L. If $\gamma \in L$, then

$$l_{v+\gamma} = l_v + \gamma.$$

Thus the twistor lines in $\tilde{Z}$ define *twistor lines* in Z, namely holomorphic sections with normal bundle isomorphic to $\mathbb{C}^{2k} \otimes \mathcal{O}(1)$.

In summary we obtain the theorem of Hitchin-Karlhede-Lindström-Roček [**HKLR**]:

THEOREM 6.7.1. *The twistor space Z is a complex manifold with holomorphic fibration $Z \xrightarrow{\Pi_Z} \mathbb{P}^1$. Furthermore there exists a holomorphic $\Pi_Z^* \mathcal{O}(2)$-valued exterior 2-form Ω on Z inducing a nondegenerate complex-symplectic structure on each fiber of Π_Z. There exists a family of holomorphic sections l_v of Π_Z each with normal bundle isomorphic to $\mathbb{C}^{2k} \otimes \mathcal{O}(1)$. There exists a real structure $\widetilde{\alpha}$ on Z covering the antipodal map $\mathbb{P}^1 \xrightarrow{\alpha} \mathbb{P}^1$, preserving each twistor line l_v and $(\widetilde{\alpha})^* \Omega = -\overline{\Omega}$.*

In terms of the coordinate atlas, L acts on $\tilde{Z}$ as follows. On the fiber $\tilde{Z}_0$, the action of $\gamma \in L$ is:

$$(6.7.1) \qquad F^{(1)}\left(\begin{bmatrix} \mathsf{q} \\ \mathsf{p} \end{bmatrix}, \psi^{(1)}(0)\right) \xmapsto{\gamma} F^{(1)}\left(\begin{bmatrix} \mathsf{q} + \gamma \\ \mathsf{p} \end{bmatrix}, \psi^{(1)}(0)\right).$$

Applying (6.5.1), the action of $\gamma \in L$ on the opposite fiber $\tilde{Z}_\infty$ is:

$$F^{(2)}\left(\begin{bmatrix} \mathsf{q} \\ \mathsf{p} \end{bmatrix}, \psi^{(2)}(0)\right) \xmapsto{\gamma} F^{(2)}\left(\begin{bmatrix} \mathsf{q} \\ \mathsf{p} - i\bar{\gamma} \end{bmatrix}, \psi^{(2)}(0)\right)$$

Applying (6.4.1) and (6.4.2), the action of $\gamma \in L$ on the other fibers are:

$$F^{(1)}\left(\begin{bmatrix} \mathsf{q} \\ \mathsf{p} \end{bmatrix}, \psi^{(1)}(\zeta)\right) \xmapsto{\gamma} F^{(1)}\left(\begin{bmatrix} \mathsf{q} + \gamma \\ \mathsf{p} - i\zeta\bar{\gamma} \end{bmatrix}, \psi^{(1)}(\zeta)\right)$$

$$F^{(2)}\left(\begin{bmatrix} \mathsf{q} \\ \mathsf{p} \end{bmatrix}, \psi^{(2)}(\xi)\right) \xmapsto{\gamma} F^{(2)}\left(\begin{bmatrix} \mathsf{q} + \xi\gamma \\ \mathsf{p} - i\bar{\gamma} \end{bmatrix}, \psi^{(2)}(\xi)\right)$$

6.8. Functions and flows. The holomorphic $\mathbb{C}^*$-action on the fiber $Z_0 = T^* \mathsf{Jac}(X)$ in §3 extends to a holomorphic $\mathbb{C}^*$-action on Z. This $\mathbb{C}^*$-action covers a holomorphic $\mathbb{C}^*$-action on $\mathbb{P}^1$, where the action of $\lambda \in \mathbb{C}^*$ is defined by:

$$\psi^{(1)}(\zeta) \longmapsto \psi^{(1)}(\lambda\zeta)$$
$$\psi^{(2)}(\xi) \longmapsto \psi^{(2)}(\lambda^{-1}\xi).$$

The $\mathbb{C}^*$-action on Z preserves the foliation by twistor lines. Since the fiber Z_0 is a cross-section for this foliation, (6.4.1) implies:

PROPOSITION 6.8.1. *The $\mathbb{C}^*$-action is described as:*

$$F^{(1)}\left(\begin{bmatrix} \mathsf{q} \\ \mathsf{p} \end{bmatrix}, \psi^{(1)}(\zeta)\right) \xmapsto{\lambda} F^{(1)}\left(\begin{bmatrix} \frac{1+|\lambda\zeta|^2}{1+|\zeta|^2} \mathsf{q} + i\zeta \frac{|\lambda|^2-1}{1+|\zeta|^2} \bar{\mathsf{p}} \\ \lambda\mathsf{p} \end{bmatrix}, \psi^{(1)}(\lambda\zeta)\right)$$

in $\mathbb{A}_1$ and

$$F^{(2)}\left(\begin{bmatrix} \mathsf{q} \\ \mathsf{p} \end{bmatrix}, \psi^{(2)}(\xi)\right) \xmapsto{\lambda} F^{(2)}\left(\begin{bmatrix} \lambda^{-1} \frac{|\lambda|^2+|\xi|^2}{1+|\xi|^2} \mathsf{q} + i\lambda^{-1}\xi \frac{|\lambda|^2-1}{1+|\xi|^2} \bar{\mathsf{p}} \\ \mathsf{p} \end{bmatrix}, \psi^{(2)}(\lambda^{-1}\xi)\right)$$

in $\mathbb{A}_2$. In particular, if $|\lambda| = 1$, then

$$F^{(1)}\left(\begin{bmatrix}\mathsf{q}\\\mathsf{p}\end{bmatrix}, \psi^{(1)}(\zeta)\right) \xmapsto{\lambda} F^{(1)}\left(\begin{bmatrix}\mathsf{q}\\\lambda\mathsf{p}\end{bmatrix}, \psi^{(1)}(\lambda\zeta)\right).$$

PROOF. The point represented by

$$F^{(1)}\left(\begin{bmatrix}\mathsf{q}\\\mathsf{p}\end{bmatrix}, \psi^{(1)}(\zeta)\right) \in \Pi_{\tilde{Z}}^{-1}(\mathbb{A}_1)$$

lies on the same twistor line as

$$F^{(1)}\left(\frac{1}{1+|\zeta|^2}\begin{bmatrix}\mathsf{q} - i\zeta\bar{\mathsf{p}}\\\mathsf{p} + i\zeta\bar{\mathsf{q}}\end{bmatrix}, \psi^{(1)}(0)\right)$$

which maps under $\lambda \in \mathbb{C}^*$ to $F^{(1)}(\mathsf{v}_\lambda, \psi^{(1)}(0))$, where

$$\mathsf{v}_\lambda := \begin{bmatrix}\mathsf{q}_\lambda\\\mathsf{p}_\lambda\end{bmatrix} := \frac{1}{1+|\zeta|^2}\begin{bmatrix}\mathsf{q} - i\zeta\bar{\mathsf{p}}\\\lambda(\mathsf{p} + i\zeta\bar{\mathsf{q}})\end{bmatrix}.$$

The twistor line containing $F^{(1)}(\mathsf{v}_\lambda, \psi^{(1)}(0))$ intersects the the fiber $Z_{\lambda\zeta}$ at

$$F^{(1)}\left(\begin{bmatrix}\mathsf{q}_\lambda + i(\lambda\zeta)\bar{\mathsf{p}}_\lambda\\\mathsf{p}_\lambda - i(\lambda\zeta)\bar{\mathsf{q}}_\lambda\end{bmatrix}, \psi^{(1)}(\lambda\zeta)\right).$$

The other claims follow similarly. $\square$

As $\lambda \longrightarrow 0$, the orbit converges to a point in $\mathcal{M}$ which is fixed by the $\mathbb{C}^*$-action and hence lies in the critical set of the Hitchin map (3.3.3).

The circle action inside the $\mathbb{C}^*$-action is Hamiltonian for the symplectic structure ω_I. The Hamiltonian potential function is the energy function defined in (3.3.5):

$$\mathbb{H}^k/L \xrightarrow{\phi} \mathbb{R}$$

$$[\mathsf{q} + \mathsf{p}J] \longmapsto \frac{1}{2}\mathsf{p} \cdot \bar{\mathsf{p}}.$$

Compare (3.3.6). In contrast:

THEOREM 6.8.2. *The function ϕ is a Kähler potential for the metric g and the complex structure*

$$e^{I\theta}J = \cos(\theta)J + \sin(\theta)K$$

for each $\theta \in \mathbb{R}$.

Let (M, g, J) be a Kähler manifold, that is, M is a manifold with complex structure J and Riemannian metric g which is Kählerian with

respect to J. Recall that a *Kähler potential* is a function $M \xrightarrow{f} \mathbb{R}$ such that
$$-\frac{1}{2} d(df \circ J) = \omega_J.$$
Here df denotes the exterior derivative and $df \circ J$ denotes the 1-form (essentially the Hodge $\star$-operator applied to df) obtained by composition
$$TM \xrightarrow{J} TM \xrightarrow{df} \mathbb{R}.$$
The notation ω_J was introduced in (5.2.2). The proof of Theorem 6.8.2 divides into several lemmas:

LEMMA 6.8.3. *For every $u \in S^2$, the function*
$$\mathbb{H}^k \xrightarrow{\rho} \mathbb{R}$$
$$\mathsf{q} + \mathsf{p} J \longmapsto \frac{1}{2}(\mathsf{q} \cdot \bar{\mathsf{q}} + \mathsf{p} \cdot \bar{\mathsf{p}})$$
is a Kähler potential for the Riemannian metric
$$g = d\mathsf{q} \cdot d\bar{\mathsf{q}} + d\mathsf{p} \cdot d\bar{\mathsf{p}}$$
on $\mathbb{H}^k$.

PROOF. The rotation group SO(3) acts transitively on S^2 and isometrically on $(\mathbb{H}^k, g)$. Thus, by symmetry we may assume that $u = I$. Now
$$\rho = \frac{1}{2}(\mathsf{q} \cdot \bar{\mathsf{q}} + \mathsf{p} \cdot \bar{\mathsf{p}})$$
$$d\rho = \frac{1}{2}(\mathsf{q} \cdot d\bar{\mathsf{q}} + \bar{\mathsf{q}} \cdot d\mathsf{q} + \mathsf{p} \cdot d\bar{\mathsf{p}} + \bar{\mathsf{p}} \cdot d\mathsf{p})$$
$$d\rho \circ I = \frac{1}{2}(-i\mathsf{q} \cdot d\bar{\mathsf{q}} + i\bar{\mathsf{q}} \cdot d\mathsf{q} - i\mathsf{p} \cdot d\bar{\mathsf{p}} + i\bar{\mathsf{p}} \cdot d\mathsf{p})$$
$$= \frac{i}{2}(-\mathsf{q} \cdot d\bar{\mathsf{q}} + \bar{\mathsf{q}} \cdot d\mathsf{q} - \mathsf{p} \cdot d\bar{\mathsf{p}} + \bar{\mathsf{p}} \cdot d\mathsf{p})$$
and
$$d(d\rho \circ I) = \frac{i}{2}(-d\mathsf{q} \cdot d\bar{\mathsf{q}} + d\bar{\mathsf{q}} \cdot d\mathsf{q} - d\mathsf{p} \cdot d\bar{\mathsf{p}} + d\bar{\mathsf{p}} \cdot d\mathsf{p})$$
$$= \omega_I$$
as claimed. $\square$

If u is a complex structure, then a function $M \xrightarrow{\psi} \mathbb{R}$ is *u-pluriharmonic* if $d(d\psi \circ u) = 0$. The sum of a Kähler potential with a pluriharmonic function is again a Kähler potential. Theorem 6.8.2 follows from adding to ρ (the Kähler potential of Lemma 6.8.3) the pluriharmonic function $-\psi$ defined in the following lemma:

LEMMA 6.8.4. *For each $\theta \in \mathbb{R}$, the function*
$$\mathbb{H}^k/L \xrightarrow{\psi} \mathbb{R}$$
$$\mathsf{q} + \mathsf{p}J \longmapsto \mathsf{q} \cdot \bar{\mathsf{q}} - \mathsf{p} \cdot \bar{\mathsf{p}}$$
is $e^{I\theta}J$-pluriharmonic.

PROOF. From the definition of J,
$$d\mathsf{p} \circ J = d\bar{\mathsf{q}}$$
$$d\mathsf{q} \circ J = -d\bar{\mathsf{p}}$$
$$d\bar{\mathsf{p}} \circ J = -d\mathsf{q}$$
$$d\bar{\mathsf{q}} \circ J = d\mathsf{p}.$$

Now
$$\psi = \mathsf{q} \cdot \bar{\mathsf{q}} - \mathsf{p} \cdot \bar{\mathsf{p}}$$
$$d\psi = \mathsf{q} \cdot d\bar{\mathsf{q}} + \bar{\mathsf{q}} \cdot d\mathsf{q} - \mathsf{p} \cdot d\bar{\mathsf{p}} - \bar{\mathsf{p}} \cdot d\mathsf{p}$$
$$d\rho \circ e^{I\theta} = e^{-i\theta}\mathsf{q} \cdot d\bar{\mathsf{q}} + e^{i\theta}\bar{\mathsf{q}} \cdot d\mathsf{q} - e^{-i\theta}\mathsf{p} \cdot d\bar{\mathsf{p}} - e^{i\theta}\bar{\mathsf{p}} \cdot d\mathsf{p}$$
$$d\rho \circ e^{I\theta} \circ J = e^{-i\theta}\mathsf{q} \cdot d\mathsf{p} + e^{i\theta}\bar{\mathsf{q}} \cdot (-d\bar{\mathsf{p}}) - e^{-i\theta}\mathsf{p} \cdot (-d\mathsf{q}) - e^{i\theta}\bar{\mathsf{p}} \cdot d\bar{\mathsf{q}}$$
$$= e^{-i\theta}\mathsf{q} \cdot d\mathsf{p} - e^{i\theta}\bar{\mathsf{q}} \cdot d\bar{\mathsf{p}} + e^{-i\theta}\mathsf{p} \cdot d\mathsf{q} - e^{i\theta}\bar{\mathsf{p}} \cdot d\bar{\mathsf{q}}$$
$$= d(e^{-i\theta}\mathsf{p} \cdot \mathsf{q} - e^{i\theta}\bar{\mathsf{p}} \cdot \bar{\mathsf{q}})$$
is closed, that is, $d(d\rho \circ (e^{I\theta}J)) = 0$, as claimed. □

The function ϕ is *not* a Kähler potential for any of the other complex structures in the hyperkähler family.

7. The moduli space and the Riemann period matrix

Here we explicitly compute these moduli spaces in terms of classical invariants. We return to the original point of view, involving the fundamental group and the Betti moduli space.

7.1. Coordinates for the Betti moduli space.
Under the direct product decomposition of Betti moduli space
$$\mathsf{Hom}(\pi, \mathbb{C}^*) \xrightarrow{\cong} \mathsf{Hom}(\pi, \mathsf{U}(1)) \times \mathsf{Hom}(\pi, \mathbb{R}^+)$$
$$\rho \longmapsto (\rho_u, \rho_\mathbb{R}),$$
the representation ρ decomposes into a unitary part ρ_u and a positive real part $\rho_\mathbb{R}$ as:
$$\rho_u(\gamma) = \frac{\rho(\gamma)}{|\rho(\gamma)|}, \qquad \rho_\mathbb{R}(\gamma) = |\rho(\gamma)|.$$

By the real analytic isomorphism $\mathbb{R}^+ \xrightarrow{\log} \mathbb{R}$, we identify
$$\mathsf{Hom}(\pi, \mathbb{R}^+) \longrightarrow \mathsf{Hom}(\pi, \mathbb{R}) \longrightarrow \mathrm{H}^1(X, \mathbb{R}) \cong \mathbb{R}^{2k}$$

Let $A_1, B_1, \ldots, A_k, B_k \in \pi$ be generators for the presentation (2.1.1). In terms of these generators, a representation $\rho \in \mathsf{Hom}(\pi, \mathbb{C}^*)$ is determined by an arbitrary $2k$-tuple
$$\bigl(\rho(A_1), \ldots, \rho(B_k)\bigr) \in (\mathbb{C}^*)^{2k}.$$
The unitary part ρ_u is then given by
$$\bigl(\frac{\rho(A_1)}{|\rho(A_1)|}, \ldots, \frac{\rho(B_k)}{|\rho(B_k)|}\bigr) \in \mathrm{U}(1)^{2k}$$
and the positive-real part by
$$\bigl(|\rho(A_1)|, \ldots, |\rho(B_k)|\bigr) \in (\mathbb{R}^+)^{2k}$$
which we identify with the real vector
$$\bigl(\log|\rho(A_1)|, \ldots, \log|\rho(B_k)|\bigr) \in \mathbb{R}^{2k}.$$

With a conformal structure on X, ρ_u corresponds to a point in $\mathsf{Jac}(X) = V/L$, where $V \cong \mathbb{C}^k$ is a complex vector space and L is the lattice in $\mathbb{C}^k$ spanned by $\mathbb{Z}^k$ and the columns of the period matrix Π. The real part $\rho_\mathbb{R}$ corresponds to a point in the real symplectic vector space $\mathrm{H}^1(X; \mathbb{R}) \cong \mathbb{R}^{2k}$. Thus, the conformal structure provides a linear $\mathbb{C}^*$-action generated by a complex structure $\mathbb{J}_\Pi$ depending on Π.

7.2. Abelian differentials and their periods. The homology classes corresponding to the generators $A_1, B_1 \ldots, A_k, B_k \in \pi$ determine a symplectic basis (also denoted by $A_1, B_1, \ldots A_k, B_k$) of $\mathrm{H}_1(X; \mathbb{Z})$ satisfying:
$$A_j \cdot A_l = A_j \cdot A_l = 0$$
$$A_j \cdot B_l = -B_j \cdot A_l = \delta_{j,l}$$
with respect to the intersection form on $\mathrm{H}_1(X; \mathbb{Z})$. If ϕ, ψ are 1-forms on X, then

(7.2.1) $$\int_X \phi \wedge \psi = \sum_{j=1}^{k} \left(\int_{A_j} \phi \int_{B_j} \psi - \int_{B_j} \phi \int_{A_j} \psi \right).$$

Recall that the space $\bar{V} = \mathcal{H}^{1,0}(X)$ of abelian differentials is a complex vector space for which the Hermitian inner product defined by (3.3.1) is negative definite. Choose a basis $\{\omega_1, \ldots, \omega_k\}$ of abelian differentials. Call such a basis *unitary* if the corresponding basis $\{\bar{\omega}_1, \ldots, \bar{\omega}_k\}$ of V is orthonormal with respect to the *positive definite* Hermitian inner product on V induced by (3.3.1).

7. THE MODULI SPACE AND THE RIEMANN PERIOD MATRIX

We record this basis as a column vector of 1-forms

$$\omega = \begin{bmatrix} \omega_1 \\ \vdots \\ \omega_k \end{bmatrix}.$$

If $\{\omega'_1, \ldots, \omega'_k\}$ is another basis, then the corresponding vectors ω and ω' satisfy

$$\omega' = U\omega,$$

where $U \in \mathrm{GL}(k, \mathbb{C})$ is an invertible $k \times k$ matrix. If ω is a unitary basis, then ω' is unitary if and only if $U \in \mathrm{U}(k)$ is a unitary matrix.

If $\alpha, \beta \in \bar{V}$ are abelian differentials, then $\star \alpha = i\alpha$ and $\star \bar{\beta} = -i\beta$, and (3.3.1) implies

$$\langle \alpha, \beta \rangle = \int_X \alpha \wedge \bar{\beta}$$

The *period matrix* for ω is defined as the $k \times 2k$ matrix $\begin{bmatrix} \mathbf{A} & \mathbf{B} \end{bmatrix}$, where the $k \times k$-matrices $\mathbf{A}, \mathbf{B}$ are defined by:

(7.2.2) $$\mathbf{A}_{j,l} := \int_{A_l} \omega_j, \qquad \mathbf{B}_{j,l} := \int_{B_l} \omega_j$$

respectively. Formula (7.2.1) implies that $\mathbf{A}$ and $\mathbf{B}$ are invertible. If ω' is another basis related to ω by $\omega' = U\omega$, then the period matrix $\begin{bmatrix} \mathbf{A}' & \mathbf{B}' \end{bmatrix}$ of ω' relates to the period matrix of ω by:

(7.2.3) $$\begin{bmatrix} \mathbf{A}' & \mathbf{B}' \end{bmatrix} = U \begin{bmatrix} \mathbf{A} & \mathbf{B} \end{bmatrix}.$$

Since the $(2,0)$-form $\omega_j \wedge \omega_l$ is identically zero,

$$0 = \int_X \omega_j \wedge \omega_l$$

and (7.2.1) implies:

(7.2.4) $$0 = \mathbf{A}\mathbf{B}^\dagger - \mathbf{B}\mathbf{A}^\dagger.$$

Suppose that ω is a unitary basis. Then

$$-i\,\delta_{j,l} = \int_X \omega_j \wedge \bar{\omega}_l$$

and (7.2.1) implies

(7.2.5) $$-i\mathbb{I} = \mathbf{A}\bar{\mathbf{B}}^\dagger - \mathbf{B}\bar{\mathbf{A}}^\dagger.$$

The *normalized basis of abelian differentials* $\{\omega'_1, \ldots, \omega'_k\}$ is the basis such that

$$\int_{A_j} \omega'_l = \delta_{j,l}.$$

Together with (7.2.3), this implies:
$$\omega' = \mathbf{A}^{-1}\omega,$$
where $\mathbf{A}$ is defined in (7.2.2). By (7.2.3), the corresponding *period matrix* is
$$\begin{bmatrix} \mathbf{A}' & \mathbf{B}' \end{bmatrix} = \begin{bmatrix} \mathbb{I} & \Pi \end{bmatrix}$$
where $\Pi = \mathbf{A}^{-1}\mathbf{B}$ is the matrix
$$\Pi_{i,j} := \int_{B_j} \omega_i'.$$
(7.2.4) and (7.2.5) imply the *Riemann bilinear relations*:
$$\Pi = \Pi^\dagger$$
$$\operatorname{Im}(\Pi) = \frac{\Pi - \bar{\Pi}}{2i} = \frac{1}{2}(\mathbf{A}^{-1})\overline{(\mathbf{A}^{-1})}^\dagger > 0.$$

That is, Π is a $k \times k$ symmetric matrix whose imaginary part is positive definite.

For a symmetric $k \times k$ complex matrix Π with positive definite imaginary part, the matrix $\begin{bmatrix} \mathbf{A} & \mathbf{B} \end{bmatrix}$ corresponding to a unitary basis can be recovered. Namely, let $\sqrt{2\,\operatorname{Im}(\Pi)}$ denote the unique positive definite symmetric square-root of $2\operatorname{Im}(\Pi)$ and let $U \in \mathrm{U}(k)$ be an arbitrary unitary matrix. Then
$$\mathbf{A} := U\bigl(\sqrt{2\,\operatorname{Im}(\Pi)}\bigr)^{-1}$$
$$\mathbf{B} := U\bigl(\sqrt{2\,\operatorname{Im}(\Pi)}\bigr)^{-1}\Pi$$
define a period matrix $\begin{bmatrix} \mathbf{A} & \mathbf{B} \end{bmatrix}$ which is unique up to the choice of U.

7.3. Flat connections. Consider a flat connection D on a trivial line bundle over X. Write
$$D = D_0 + \psi + \phi,$$
where D_0 is the flat connection arising from the trivialization, ψ purely imaginary and ϕ real 1-forms. By a gauge transformation, one may write
$$\psi = \Psi - \bar{\Psi}$$
$$\phi = \Phi + \bar{\Phi},$$
where $\Psi \in V$ and $\Phi \in \bar{V}$. The corresponding vectors $\mathsf{q}, \mathsf{p} \in \mathbb{C}^k$ satisfy
$$\Psi = \sum_{j=1}^k q_j \bar{\omega}_j = \bar{\omega}^\dagger \cdot \mathsf{q}, \qquad \Phi = \sum_{j=1}^k p_j \omega_j = \omega^\dagger \cdot \mathsf{p},$$

where
$$\mathbf{q} = \begin{bmatrix} q_1 \\ \vdots \\ q_k \end{bmatrix}, \qquad \mathbf{p} = \begin{bmatrix} p_1 \\ \vdots \\ p_k \end{bmatrix}.$$

The Higgs pair consists of the holomorphic structure $D'' = D_0'' + \mathbf{q}^\dagger \bar{\omega}$ and the Higgs field $\Phi = \mathbf{p}^\dagger \omega$. Corresponding to $\mathbf{q}, \mathbf{p} \in \mathbb{C}^k$ is the connection $D = D_0 + \eta$, where
$$\eta = \left(\mathbf{q}^\dagger \bar{\omega} - \bar{\mathbf{q}}^\dagger \omega\right) + \left(\bar{\omega}^\dagger \cdot \mathbf{q} - \omega^\dagger \cdot \bar{\mathbf{q}}\right) + \left(\omega^\dagger \cdot \mathbf{p} + \bar{\omega}^\dagger \cdot \bar{\mathbf{p}}\right).$$

To simplify calculations, henceforth let ω be the normalized basis of abelian differentials such that $\mathbf{A} = \mathbb{I}$ and $\mathbf{B} = \Pi$.

If D is a flat unitary connection, then $D = D_0 + \psi$, where ψ is purely imaginary. By a gauge transformation, one may assume that $\psi = \Psi - \bar{\Psi}$, where $\Psi \in \mathcal{H}^{0,1}(X)$. The corresponding vector $\mathbf{q} \in \mathbb{C}^k$ is defined by
$$\Psi = \sum_{j=1}^{k} q_j \bar{\omega}_j = \bar{\omega}^\dagger \cdot \mathbf{q}$$

and
$$\psi = \bar{\omega}^\dagger \cdot \mathbf{q} - \omega^\dagger \cdot \bar{\mathbf{q}} = 2i \operatorname{Im}\left(\bar{\omega}^\dagger \cdot \mathbf{q}\right),$$

with corresponding periods
$$\int_{A_l} \psi = 2i \operatorname{Im}(q_l)$$
$$\int_{B_l} \psi = 2i \operatorname{Im}\left(\sum_{j=1}^{k} q_j \bar{\Pi}_{j,l}\right).$$

In the notation
$$\mathbf{a} := \begin{bmatrix} \int_{A_1} \psi \\ \vdots \\ \int_{A_k} \psi \end{bmatrix}, \qquad \mathbf{b} := \begin{bmatrix} \int_{B_1} \psi \\ \vdots \\ \int_{B_k} \psi \end{bmatrix},$$

the periods are:

$\mathbf{a} = 2i \operatorname{Im}(\mathbf{q})$

$\mathbf{b} = 2i \operatorname{Im}(\bar{\Pi} \cdot \mathbf{q}) = 2i \left(\operatorname{Re}(\Pi) \cdot \operatorname{Im}(\mathbf{q}) - \operatorname{Im}(\Pi) \cdot \operatorname{Re}(\mathbf{q})\right).$

The lattice L in $\mathbb{C}^k$ consists of $\mathbf{q}$ such that the periods $\mathbf{a}, \mathbf{b} \in 2\pi i \mathbb{Z}$, that is,
$$\operatorname{Im}(\mathbf{q}) \in \pi \mathbb{Z}^k, \qquad \operatorname{Re}(\Pi) \operatorname{Im}(\mathbf{q}) - \operatorname{Im}(\Pi) \operatorname{Re}(\mathbf{q}) \in \pi \mathbb{Z}^k$$
which is equivalent to:
$$\mathbf{q} \in \pi \operatorname{Im}(\Pi)^{-1}\{\mathbb{Z}^k + \Pi \mathbb{Z}^k\}.$$

Thus L is equivalent (by $\pi\,\mathrm{Im}(\Pi)^{-1} \in \mathrm{GL}(k,\mathbb{C})$) to the lattice $\mathbb{Z}^k + \Pi\mathbb{Z}^k$ spanned by the $2k$ columns of $\begin{bmatrix} \mathbb{I}_k & \Pi \end{bmatrix}$. Therefore $\mathsf{Jac}(X)$ is the quotient of $\mathbb{C}^k$ by $\mathbb{Z}^k + \Pi\mathbb{Z}^k$. (See, for example, Gunning [**Gu1, Gu3**], §2.5.)

7.4. Higgs fields. The space $\mathsf{Hom}(\pi, \mathbb{R}^+)$ of positive real characters inherits structure from the conformal structure of X through the period matrix Π. For $\rho \in \mathsf{Hom}(\pi, \mathbb{C}^*)$, write

$$\alpha_j := \frac{1}{2}\log|\rho(A_j)|, \qquad \beta_j := \frac{1}{2}\log|\rho(B_j)|.$$

We determine the corresponding flat $\mathbb{R}^+$-connection with holonomy (α, β).

After applying a gauge transformation, the 1-form $\phi \in \mathcal{H}^1(X;\mathbb{R})$ decomposes as $\phi = \Phi + \bar{\Phi}$, where $\Phi \in \mathcal{H}^{1,0}(X)$ is a holomorphic 1-form. The vector $\mathsf{p} \in \mathbb{C}^k$ corresponding to Φ is defined by

$$\Phi = \sum_{j=1}^{k} p_j \omega_j = \omega^\dagger \mathsf{p}$$

which has periods

$$\alpha_j = \frac{1}{2}\int_{A_j} \phi = \mathrm{Re}\int_{A_j} \Phi = \mathrm{Re}(p_j)$$

$$\beta_j = \frac{1}{2}\int_{B_j} \phi = \mathrm{Re}\int_{B_j} \Phi = \mathrm{Re}\Big(\sum_{l=1}^{k} \Pi_{j,l} p_l\Big).$$

Thus

$$\alpha = \mathrm{Re}(\mathsf{p})$$
$$\beta = \mathrm{Re}(\Pi)\,\mathrm{Re}(\mathsf{p}) - \mathrm{Im}(\Pi)\,\mathrm{Im}(\mathsf{p})$$

with inverse mapping

$$\mathrm{Re}(\mathsf{p}) = \alpha$$
$$\mathrm{Im}(\mathsf{p}) = \mathrm{Im}(\Pi)^{-1}\big(\mathrm{Re}(\Pi)\alpha - \beta\big).$$

As matrices,

$$\begin{bmatrix} \mathrm{Re}(\mathsf{p}) \\ \mathrm{Im}(\mathsf{p}) \end{bmatrix} \longmapsto \begin{bmatrix} \mathbb{I}_k & 0 \\ \mathrm{Re}(\Pi) & -\mathrm{Im}(\Pi) \end{bmatrix} \begin{bmatrix} \alpha \\ \beta \end{bmatrix}$$

and

$$\begin{bmatrix} \alpha \\ \beta \end{bmatrix} \longmapsto \begin{bmatrix} \mathbb{I}_k & 0 \\ \mathrm{Im}(\Pi)^{-1}\mathrm{Re}(\Pi) & -\mathrm{Im}(\Pi)^{-1} \end{bmatrix} \begin{bmatrix} \mathrm{Re}(\mathsf{p}) \\ \mathrm{Im}(\mathsf{p}) \end{bmatrix}.$$

7. THE MODULI SPACE AND THE RIEMANN PERIOD MATRIX

7.5. The $\mathbb{C}^*$-action in terms of the period matrix. We describe the $\mathbb{C}^*$-action explicitly as follows. We break p into its real and imaginary parts:
$$\mathsf{p} = \operatorname{Re}(\mathsf{p}) + i\operatorname{Im}(\mathsf{p})$$
Scalar multiplication $\mathsf{p} \longmapsto i\mathsf{p}$ is:
$$\begin{bmatrix}\operatorname{Re}(\mathsf{p})\\ \operatorname{Im}(\mathsf{p})\end{bmatrix} \longmapsto \begin{bmatrix}\operatorname{Im}(\mathsf{p})\\ -\operatorname{Re}(\mathsf{p})\end{bmatrix} = \begin{bmatrix}0 & \mathbb{I}_k \\ -\mathbb{I}_k & 0\end{bmatrix}\begin{bmatrix}\operatorname{Re}(\mathsf{p})\\ \operatorname{Im}(\mathsf{p})\end{bmatrix}.$$
Multiplying the Higgs field Φ by i transforms the periods α, β by:
$$\begin{bmatrix}\alpha\\ \beta\end{bmatrix} \longmapsto \mathbb{J}_\Pi \begin{bmatrix}\alpha\\ \beta\end{bmatrix}$$
where
$$\mathbb{J}_\Pi = \begin{bmatrix}\mathbb{I}_k & 0 \\ \operatorname{Re}(\Pi) & -\operatorname{Im}(\Pi)\end{bmatrix}\begin{bmatrix}0 & \mathbb{I}_k \\ -\mathbb{I}_k & 0\end{bmatrix}\begin{bmatrix}\mathbb{I}_k & 0 \\ \operatorname{Im}(\Pi)^{-1}\operatorname{Re}(\Pi) & -\operatorname{Im}(\Pi)^{-1}\end{bmatrix}$$

(7.5.1)
$$= \begin{bmatrix}-\operatorname{Im}(\Pi)^{-1}\operatorname{Re}(\Pi) & \operatorname{Im}(\Pi)^{-1} \\ -\operatorname{Re}(\Pi)\operatorname{Im}(\Pi)^{-1}\operatorname{Re}(\Pi) - \operatorname{Im}(\Pi) & \operatorname{Re}(\Pi)\operatorname{Im}(\Pi)^{-1}\end{bmatrix}.$$
Then the $\mathbb{C}^*$-action is given by
$$\operatorname{Re}(\lambda)\mathbb{I}_{2k} + \operatorname{Im}(\lambda)\mathbb{J}_\Pi$$
for $\lambda \in \mathbb{C}^*$.

7.6. The $\mathbb{C}^*$-action and the real points. The $\mathbb{C}^*$-action on $\mathsf{Hom}(\pi, \mathbb{C}^*)$ fixes the unitary component and acts as above on the Higgs field. That is, $\lambda \in \mathbb{C}^*$ maps
$$\rho = (\rho_u, \rho_\mathbb{R}) \longmapsto \Big(\rho_u, \exp\big((\operatorname{Re}(\lambda)\mathbb{I} + \operatorname{Im}(\lambda)\mathbb{J}_\Pi)\log(\rho_\mathbb{R})\big)\Big),$$
where $\mathbb{J}_\Pi$ is defined by (7.5.1).

The unitary projection $\mathsf{Hom}(\pi, \mathbb{C}^*) \longrightarrow \mathsf{Hom}(\pi, \mathsf{U}(1))$ associates to a Higgs line bundle the underlying holomorphic line bundle. It can be described purely in terms of the $\mathbb{C}^*$-action as follows. Let $\rho \in \mathsf{Hom}(\pi, \mathbb{C}^*)$. Then as $\lambda \longrightarrow 0$, the orbit $\lambda \cdot \rho$ approaches ρ_u. In particular this structure defines the foliation of $\mathsf{Hom}(\pi, \mathbb{C}^*)$ by copies of $\mathsf{Hom}(\pi, \mathbb{R}^+) = \mathsf{H}^1(X; \mathbb{R})$.

Moreover the complex-symplectic structure (J, Ω) defined by (2.2.7) arises from the complex-orthogonal structure on $\mathbb{C}^*$. By (2.2.9) (compare also (5.3.2)), the imaginary part of Ω is a real-symplectic structure $-\omega_K$ under which the decomposition
$$\mathsf{Hom}(\pi, \mathbb{C}^*) = \mathsf{Hom}(\pi, \mathsf{U}(1)) \times \mathsf{Hom}(\pi, \mathbb{R}^+)$$

is a Lagrangian product (a *real bipolarization*). The $\mathbb{C}^*$-action is the extra piece of structure to determine the Dolbeault moduli space (expressed in terms of the complex structure I). As $\mathsf{Hom}(\pi, \mathrm{U}(1))$ is ω_K-Lagrangian, ω_K identifies the tangent space with the normal space, which in turn identifies with $\mathrm{H}^1(X;\mathbb{R})$. This real-symplectic space inherits a Hermitian structure from the complex structure $\mathbb{J}_\Pi$, which defines a complex structure on $\mathsf{Hom}(\pi, \mathrm{U}(1))$. With this complex structure, $\mathsf{Hom}(\pi, \mathrm{U}(1))$ identifies with $\mathsf{Jac}(X)$. Thus the full structure of the Dolbeault moduli space of Higgs bundles arises from the dynamics of the $\mathbb{C}^*$-action and the natural structures of the Betti moduli space which depend only on the fundamental group of Σ.

Bibliography

[ABCKT] Amorós, J., Burger, M., Corlette, K., Kotschick, D. and Toledo, D., *Fundamental Groups of Compact Kähler manifolds*, Mathematical Surveys and Monographs vol 44, Amer. Math. Soc., Providence, RI, 1996.

[A] Arapura, D., *Higgs line bundles, Green-Lazarsfeld sets, and maps of Kähler manifolds to curves*, Bull. Amer. Math. Soc. *26* (2), (1992), 31–314.

[AB] Atiyah, M., and Bott, R., *The Yang-Mills equations over Riemann surfaces*, Phil. Trans. R. Soc. Lond. *A 308* (1982), 523–615.

[B] Besse, A., *Einstein Manifolds*, Springer-Verlag, Berlin, Heidelberg, New York, 1987.

[BGG1] Bradlow, S., Garca-Prada, O. Gothen, P., *Surface group representations and* $U(p,q)$-*Higgs bundles*, J. Diff. Geom. *64* (2003), no. 1, 111–170.

[BGG2] *Representations of the fundamental group of a surface in* $PU(p,q)$ *and holomorphic triples*, C. R. Acad. Sci. Paris Sr. I Math. *333* (2001), no. 4, 347–352.

[C1] Corlette, K., *Flat bundles with canonical metrics*, J. Diff. Geom. *28* (1988), 361–382.

[C2] _____, *Nonabelian Hodge theory*, Proc. Symp. in Pure Math. Vol 54, Part 2, Amer. Math. Soc., Providence, RI, 1993, pp. 125–144.

[D] Dancer, A., *Hyperkähler manifolds,* in "Essays on Einstein Manifolds,", Surveys in Differential Geometry VI (C. Le Brun and M. Wang, eds.), International Press, 1999, pp 15–38.

[D1] Donaldson, S., *A new proof of a theorem of Narasimhan and Seshadri*, J. Diff. Geom. *18* (1983), 269–277.

[D2] _____, *Twisted harmonic maps and the self-duality equations*, Proc. London Math. Soc. *55* (3), (1987), 127–131.

[D3] _____, *Moment maps in Differential Geometry,* (to appear).

[ES] Eells, J. and Sampson, J., *Harmonic mappings of Riemannian manifolds*, American Journal of Mathematics *86* (1964), 109–160.

[FK] Farkas, H. and Kra, I., *Riemann Surfaces*, Graduate Texts in Mathematics vol. 71, Springer-Verlag, Berlin, Heidelberg, New York 1980.

[Fo] Forster, O. *Lectures on Riemann Surfaces*, Graduate Texts in Mathematics vol. 81 Springer Verlag 1999.

[Fu] Fulton, W., *Algebraic Topology: A first course*, Graduate Texts in Mathematics vol. 153 Springer-Verlag, Berlin, Heidelberg, New York, 1995.

[GM] Gelfand, S. I. and Manin,Y., *Homological Algebra,* in "Algebra V", Encyclopedia of Mathematical Sciences, A. I. Kostrikin, I. R. Shafarevich (Eds.), Springer-Verlag New York.

[G] Goldman, W., *The symplectic nature of fundamental groups of surfaces*, Adv. Math. *54* (1984), 200–225.

[G2] _____, *Invariant functions on Lie groups and Hamiltonian flows of surface group representations*, Inv. Math. **85** (1986), 1–40.

[GoMi] _____ and Millson, J., *The deformation theory of representations of fundamental groups of compact Kähler manifolds*, Publ. Math. I.H.E.S. **67** (1988), 43–96.

[Go1] Gothen, P., *The Betti numbers of the moduli space of stable rank* 3 *Higgs bundles on a Riemann surface*, Internat. J. Math. *5* (1994), no. 6, 861–875.

[Go2] _____, *Components of spaces of representations and stable triples*, Topology *40* (2001), no. 4, 823–850.

[Go3] _____, *Topology of* $U(2,1)$ *representation spaces*, Bull. London Math. Soc. *34* (2002), no. 6, 729–738.

[Gr] Griffiths, P. and Harris, J., *Principles of Algebraic Geometry*, Wiley Interscience, 1978.

[Gu1] Gunning, R. C., *Lectures on Riemann Surfaces*, Mathematical Notes Vol. 2, Princeton University Press 1966.

[Gu2] _____, *Lectures on Vector Bundles over Riemann Surfaces*, Mathematical Notes vol 6, Princeton University Press, 1967.

[Gu3] _____, *Riemann Surfaces and Generalized Theta Functions* Ergebnisse der Mathematik und ihrere Grenzgebiete vol. 91, Springer-Verlag Berlin-Heidelberg-New York, 1976.

[H1] Hitchin, N. J., *The self-duality equations on Riemann surfaces*, Proc. London Math. Soc. (3) **55** (1987), 59–126.

[H2] _____, *Lie groups and Teichmüller space*, Topology *31* (3), (1992) 449–473.

[H3] _____, *Hyperkähler manifolds*, Séminaire Bourbaki, Vol. 1991/2 Asterisque *206* (1992), 137–166.

[H4] _____, *Stable bundles and integrable systems*, Duke Math. J. *54* No. 1 (1987), 91–114.

[HKLR] _____, Karlhede, A., Lindström, U., and Roček, M., *Hyperkähler metrics and supersymmetry*, Comm. Math. Phys. *108* (1987), 535–589.

[HSW] _____, Segal, G., and Ward, R., *Integrable Systems: Twistors, Loop Groups, and Riemann Surfaces*, Oxford Graduate Texts in Mathematics Vol. 4, Oxford University Press, 1999.

[Ja] Jacobson, N. *Basic Algebra II*, W.H. Freeman and Co., San Francisco, 1980.

[Jo] Jost, J., *Compact Riemann Surfaces*, Springer-Verlag, Berlin, Heidelberg, New York, 1997.

[Joy] Joyce, D., *Compact Manifolds with Special Holonomy*, Oxford Mathematical Monographs, Oxford University Press, 2000

[K] Kobayashi, S., *Differential Geometry of Complex Vector Bundles*, Publications of the Mathematical Society of Japan Vol. 15, Princeton University Press and Iwanami Shoten 1987.

[KN] _____ and Nomizu, K., *Foundations of Differential Geometry*, I, John Wiley & Sons, 1969.

[L] Labourie, F., *Existence d'applications harmoniques tordues à valeurs dans les variétés à courbure négative*, Proc. Amer. Math. Soc. *111*, (3), (1991), 877–882.

[MS] McDuff, D. and Salamon, D., *Introduction to Symplectic Topology*, Oxford University Press, 1998.

[GIT] Mumford, D., Fogarty, J., and Kirwan, F., *Geometric Invariant Theory*, Third Ed., Springer Verlag, 1994.

[Sa] Salamon, S., *Riemannian Geometry and Holonomy Groups*, Pittman Research Notes in Mathematics vol. 201 Longman, Harlow, 1989.

[Si1] Simpson, C., *Constructing variations of Hodge structure using Yang-Mills theory and applications to uniformization*, J. Amer. Math. Soc. *1* (1988), 867–918.

[Si2] _____, *Higgs bundles and local systems*, Publ. Math. I.H.E.S. *75* (1992), 5–95.

[Si3] _____, *Nonabelian Hodge theory*, Proc. I.C.M., Kyoto 1990, Springer-Verlag, 1991, pp. 198–230.

[Si4] _____, *The Hodge filtration in nonabelian cohomology*, in "Algebraic Geometry – Santa Cruz 1995", Proc. Symp. Pure Math. vol. 62, Part 2 Amer. Math. Soc., Providence, RI, 1997, pp. 215–281.

[Si5] _____, *Moduli of representations of the fundamental group of smooth projective variety* II, Publ. Math. I.H.E.S. *80* (1994), 5–79.

[S] Steenrod, N., *The topology of fiber bundles*, Princeton University Press, 1951.

[T] Toen, B., *Introduction to nonabelian Hodge theory* I, lecture presented at MSRI workshop, "Nonabelian Hodge theory," April 2002.

[Wei] Weinstein, A., *Lectures on symplectic manifolds*, Amer. Math. Soc., Providence, RI.

[Wel] Wells, R.O., *Differential Analysis on Complex Manifolds*, Graduate Texts in Mathematics Vol. 65, Springer-Verlag, 1980.

[Wey] Weyl, H., *The Concept of a Riemann Surface*, Addison-Wesley, 1955.

[X0] Xia, E., *Components of* $\mathsf{Hom}(\pi_1, \mathrm{PGL}(2,\mathbb{R}))$, Topology, *36*, (2), (1997) 481–499.

[X1] _____, *The moduli of flat* $\mathrm{PGL}(2,\mathbb{R})$-*connections on Riemann surfaces*, Comm. Math. Phys. *203* (1999), no. 3, 531–549.

[X2] _____, *The moduli of flat* $PU(2,1)$ *structures on Riemann surfaces*, Pacific J. Math. *195* (2000), (1), 231–256.

Editorial Information

To be published in the *Memoirs*, a paper must be correct, new, nontrivial, and significant. Further, it must be well written and of interest to a substantial number of mathematicians. Piecemeal results, such as an inconclusive step toward an unproved major theorem or a minor variation on a known result, are in general not acceptable for publication.

Papers appearing in *Memoirs* are generally at least 80 and not more than 200 published pages in length. Papers less than 80 or more than 200 published pages require the approval of the Managing Editor of the Transactions/Memoirs Editorial Board.

As of January 31, 2008, the backlog for this journal was approximately 17 volumes. This estimate is the result of dividing the number of manuscripts for this journal in the Providence office that have not yet gone to the printer on the above date by the average number of monographs per volume over the previous twelve months, reduced by the number of volumes published in four months (the time necessary for preparing a volume for the printer). (There are 6 volumes per year, each usually containing at least 4 numbers.)

A Consent to Publish and Copyright Agreement is required before a paper will be published in the *Memoirs*. After a paper is accepted for publication, the Providence office will send a Consent to Publish and Copyright Agreement to all authors of the paper. By submitting a paper to the *Memoirs*, authors certify that the results have not been submitted to nor are they under consideration for publication by another journal, conference proceedings, or similar publication.

Information for Authors

Memoirs are printed from camera copy fully prepared by the author. This means that the finished book will look exactly like the copy submitted.

Initial submission. The AMS uses Centralized Manuscript Processing for initial submissions. Authors should submit a PDF file using the Initial Manuscript Submission form found at www.ams.org/cgi-bin/peertrack/submission.pl, or send one copy of the manuscript to the following address: Centralized Manuscript Processing, MEMOIRS OF THE AMS, 201 Charles Street, Providence, RI 02904-2294 USA. If a paper copy is being forwarded to the AMS, indicate that it is for it Memoirs and include the name of the corresponding author, contact information such as email address or mailing address, and the name of an appropriate Editor to review the paper (see the list of Editors below).

The paper must contain a *descriptive title* and an *abstract* that summarizes the article in language suitable for workers in the general field (algebra, analysis, etc.). The *descriptive title* should be short, but informative; useless or vague phrases such as "some remarks about" or "concerning" should be avoided. The *abstract* should be at least one complete sentence, and at most 300 words. Included with the footnotes to the paper should be the 2000 *Mathematics Subject Classification* representing the primary and secondary subjects of the article. The classifications are accessible from www.ams.org/msc/. The list of classifications is also available in print starting with the 1999 annual index of *Mathematical Reviews*. The Mathematics Subject Classification footnote may be followed by a list of *key words and phrases* describing the subject matter of the article and taken from it. Journal abbreviations used in bibliographies are listed in the latest *Mathematical Reviews* annual index. The series abbreviations are also accessible from www.ams.org/publications/. To help in preparing and verifying references, the AMS offers MR Lookup, a Reference Tool for Linking, at www.ams.org/mrlookup/.

Electronically prepared manuscripts. The AMS encourages electronically prepared manuscripts, with a strong preference for $\mathcal{AMS}$-LaTeX. To this end, the Society has prepared $\mathcal{AMS}$-LaTeX author packages for each AMS publication. Author packages include instructions for preparing electronic manuscripts, samples, and a style file that generates

the particular design specifications of that publication series. Though $\mathcal{AMS}$-LaTeX is the highly preferred format of TeX, author packages are also available in $\mathcal{AMS}$-TeX.

Authors may retrieve an author package from the AMS website starting from www.ams.org/tex/ or via FTP to ftp.ams.org (login as anonymous, enter username as password, and type cd pub/author-info). The *AMS Author Handbook* and the *Instruction Manual* are available in PDF format following the author packages link from www.ams.org/tex/. The author package can also be obtained free of charge by sending email to tech-support@ams.org (Internet) or from the Publication Division, American Mathematical Society, 201 Charles St., Providence, RI 02904-2294, USA. When requesting an author package, please specify $\mathcal{AMS}$-LaTeX or $\mathcal{AMS}$-TeX and the publication in which your paper will appear. Please be sure to include your complete mailing address.

After acceptance. The final version of the electronic file should be sent to the Providence office (this includes any TeX source file, any graphics files, and the DVI or PostScript file) immediately after the paper has been accepted for publication.

Before sending the source file, be sure you have proofread your paper carefully. The files you send must be the EXACT files used to generate the proof copy that was accepted for publication. For all publications, authors are required to send a printed copy of their paper, which exactly matches the copy approved for publication, along with any graphics that will appear in the paper.

Accepted electronically prepared files can be submitted via the web at www.ams.org/submit-book-journal/, sent via FTP, or sent on CD-Rom or diskette to the Electronic Prepress Department, American Mathematical Society, 201 Charles Street, Providence, RI 02904-2294 USA. TeX source files, DVI files, and PostScript files can be transferred over the Internet by FTP to the Internet node ftp.ams.org (130.44.1.100). When sending a manuscript electronically via CD-Rom or diskette, please be sure to include a message identifying the paper as a Memoir.

Electronically prepared manuscripts can also be sent via email to pub-submit@ams.org (Internet). In order to send files via email, they must be encoded properly. (DVI files are binary and PostScript files tend to be very large.)

Electronic graphics. Comprehensive instructions on preparing graphics are available at www.ams.org/jourhtml/. A few of the major requirements are given here.

Submit files for graphics as EPS (Encapsulated PostScript) files. This includes graphics originated via a graphics application as well as scanned photographs or other computer-generated images. If this is not possible, TIFF files are acceptable as long as they can be opened in Adobe Photoshop or Illustrator. No matter what method was used to produce the graphic, it is necessary to provide a paper copy to the AMS.

Authors using graphics packages for the creation of electronic art should also avoid the use of any lines thinner than 0.5 points in width. Many graphics packages allow the user to specify a "hairline" for a very thin line. Hairlines often look acceptable when proofed on a typical laser printer. However, when produced on a high-resolution laser imagesetter, hairlines become nearly invisible and will be lost entirely in the final printing process.

Screens should be set to values between 15% and 85%. Screens which fall outside of this range are too light or too dark to print correctly. Variations of screens within a graphic should be no less than 10%.

Inquiries. Any inquiries concerning a paper that has been accepted for publication should be sent to memo-query@ams.org or directly to the Electronic Prepress Department, American Mathematical Society, 201 Charles St., Providence, RI 02904-2294 USA.

Editors

This journal is designed particularly for long research papers, normally at least 80 pages in length, and groups of cognate papers in pure and applied mathematics. Papers intended for publication in the *Memoirs* should be addressed to one of the following editors. The AMS uses Centralized Manuscript Processing for initial submissions to AMS journals. Authors should follow instructions listed on the Initial Submission page found at www.ams.org/memo/memosubmit.html.

Algebra to ALEXANDER KLESHCHEV, Department of Mathematics, University of Oregon, Eugene, OR 97403-1222; email: ams@noether.uoregon.edu

Algebraic geometry and its application to MINA TEICHER, Emmy Noether Research Institute for Mathematics, Bar-Ilan University, Ramat-Gan 52900, Israel; email: teicher@macs.biu.ac.il

Algebraic geometry to DAN ABRAMOVICH, Department of Mathematics, Brown University, Box 1917, Providence, RI 02912; email: amsedit@math.brown.edu

Algebraic number theory to V. KUMAR MURTY, Department of Mathematics, University of Toronto, 100 St. George Street, Toronto, ON M5S 1A1, Canada; email: murty@math.toronto.edu

Algebraic topology to ALEJANDRO ADEM, Department of Mathematics, University of British Columbia, Room 121, 1984 Mathematics Road, Vancouver, British Columbia, Canada V6T 1Z2; email: adem@math.ubc.ca

Combinatorics to JOHN R. STEMBRIDGE, Department of Mathematics, University of Michigan, Ann Arbor, Michigan 48109-1109; email: FRS@umich.edu

Complex analysis and harmonic analysis to ALEXANDER NAGEL, Department of Mathematics, University of Wisconsin, 480 Lincoln Drive, Madison, WI 53706-1313; email: nagel@math.wisc.edu

Differential geometry and global analysis to LISA C. JEFFREY, Department of Mathematics, University of Toronto, 100 St. George St., Toronto, ON Canada M5S 3G3; email: jeffrey@math.toronto.edu

Functional analysis and operator algebras to DIMITRI SHLYAKHTENKO, Department of Mathematics, University of California, Los Angeles, CA 90095; email: shlyakht@math.ucla.edu

Geometric analysis to WILLIAM P. MINICOZZI II, Department of Mathematics, Johns Hopkins University, 3400 N. Charles St., Baltimore, MD 21218; email: trans@math.jhu.edu

Geometric analysis to MARK FEIGHN, Math Department, Rutgers University, Newark, NJ 07102; email: feighn@andromeda.rutgers.edu

Harmonic analysis, representation theory, and Lie theory to ROBERT J. STANTON, Department of Mathematics, The Ohio State University, 231 West 18th Avenue, Columbus, OH 43210-1174; email: stanton@math.ohio-state.edu

Logic to STEFFEN LEMPP, Department of Mathematics, University of Wisconsin, 480 Lincoln Drive, Madison, Wisconsin 53706-1388; email: lempp@math.wisc.edu

Number theory to JONATHAN ROGAWSKI, Department of Mathematics, University of California, Los Angeles, CA 90095; email: jonr@math.ucla.edu

Partial differential equations to GUSTAVO PONCE, Department of Mathematics, South Hall, Room 6607, University of California, Santa Barbara, CA 93106; email: ponce@math.ucsb.edu

Partial differential equations and dynamical systems to PETER POLACIK, School of Mathematics, University of Minnesota, Minneapolis, MN 55455; email: polacik@math.umn.edu

Probability and statistics to RICHARD BASS, Department of Mathematics, University of Connecticut, Storrs, CT 06269-3009; email: bass@math.uconn.edu

Real analysis and partial differential equations to DANIEL TATARU, Department of Mathematics, University of California, Berkeley, Berkeley, CA 94720; email: tataru@math.berkeley.edu

All other communications to the editors should be addressed to the Managing Editor, ROBERT GURALNICK, Department of Mathematics, University of Southern California, Los Angeles, CA 90089-1113; email: guralnic@math.usc.edu.

Titles in This Series

905 **Dominic Verity,** Complicial sets characterising the simplicial nerves of strict ω-categories, 2008

904 **William M. Goldman and Eugene Z. Xia,** Rank one Higgs bundles and representations of fundamental groups of Riemann surfaces, 2008

903 **Gail Letzter,** Invariant differential operators for quantum symmetric spaces, 2008

902 **Bertrand Toën and Gabriele Vezzosi,** Homotopical algebraic geometry II: Geometric stacks and applications, 2008

901 **Ron Donagi and Tony Pantev (with an appendix by Dmitry Arinkin),** Torus fibrations, gerbes, and duality, 2008

900 **Wolfgang Bertram,** Differential geometry, Lie groups and symmetric spaces over general base fields and rings, 2008

899 **Piotr Hajłasz, Tadeusz Iwaniec, Jan Malý, and Jani Onninen,** Weakly differentiable mappings between manifolds, 2008

898 **John Rognes,** Galois extensions of structured ring spectra/Stably dualizable groups, 2008

897 **Michael I. Ganzburg,** Limit theorems of polynomial approximation with exponential weights, 2008

896 **Michael Kapovich, Bernhard Leeb, and John J. Millson,** The generalized triangle inequalities in symmetric spaces and buildings with applications to algebra, 2008

895 **Steffen Roch,** Finite sections of band-dominated operators, 2008

894 **Martin Dindoš,** Hardy spaces and potential theory on C^1 domains in Riemannian manifolds, 2008

893 **Tadeusz Iwaniec and Gaven Martin,** The Beltrami Equation, 2008

892 **Jim Agler, John Harland, and Benjamin J. Raphael,** Classical function theory, operator dilation theory, and machine computation on multiply-connected domains, 2008

891 **John H. Hubbard and Peter Papadopol,** Newton's method applied to two quadratic equations in $\mathbb{C}^2$ viewed as a global dynamical system, 2008

890 **Steven Dale Cutkosky,** Toroidalization of dominant morphisms of 3-folds, 2007

889 **Michael Sever,** Distribution solutions of nonlinear systems of conservation laws, 2007

888 **Roger Chalkley,** Basic global relative invariants for nonlinear differential equations, 2007

887 **Charlotte Wahl,** Noncommutative Maslov index and eta-forms, 2007

886 **Robert M. Guralnick and John Shareshian,** Symmetric and alternating groups as monodromy groups of Riemann surfaces I: Generic covers and covers with many branch points, 2007

885 **Jae Choon Cha,** The structure of the rational concordance group of knots, 2007

884 **Dan Haran, Moshe Jarden, and Florian Pop,** Projective group structures as absolute Galois structures with block approximation, 2007

883 **Apostolos Beligiannis and Idun Reiten,** Homological and homotopical aspects of torsion theories, 2007

882 **Lars Inge Hedberg and Yuri Netrusov,** An axiomatic approach to function spaces, spec tral synthesis and Luzin approximation, 2007

881 **Tao Mei,** Operator valued Hardy spaces, 2007

880 **Bruce C. Berndt, Geumlan Choi, Youn-Seo Choi, Heekyoung Hahn, Boon Pin Yeap, Ae Ja Yee, Hamza Yesilyurt, and Jinhee Yi,** Ramanujan's forty identities for Rogers-Ramanujan functions, 2007

879 **O. García-Prada, P. B. Gothen, and V. Muñoz,** Betti numbers of the moduli space of rank 3 parabolic Higgs bundles, 2007

878 **Alessandra Celletti and Luigi Chierchia,** KAM stability and celestial mechanics, 2007

877 **María J. Carro, José A. Raposo, and Javier Soria,** Recent developments in the theory of Lorentz spaces and weighted inequalities, 2007

TITLES IN THIS SERIES

876 **Gabriel Debs and Jean Saint Raymond,** Borel liftings of Borel sets: Some decidable and undecidable statements, 2007
875 **C. Krattenthaler and T. Rivoal,** Hypergéométrie et fonction zêta de Riemann, 2007
874 **Sonia Natale,** Semisolvability of semisimple Hopf algebras of low dimension, 2007
873 **A. J. Duncan,** Exponential genus problems in one-relator products of groups, 2007
872 **Anthony V. Geramita, Tadahito Harima, Juan C. Migliore, and Yong Su Shin,** The Hilbert function of a level algebra, 2007
871 **Pascal Auscher,** On necessary and sufficient conditions for L^p-estimates of Riesz transforms associated to elliptic operators on $\mathbb{R}^n$ and related estimates, 2007
870 **Takuro Mochizuki,** Asymptotic behaviour of tame harmonic bundles and an application to pure twistor D-modules, Part 2, 2007
869 **Takuro Mochizuki,** Asymptotic behaviour of tame harmonic bundles and an application to pure twistor D-modules, Part 1, 2007
868 **Gelu Popescu,** Entropy and multivariable interpolation, 2006
867 **Vilmos Totik,** Metric properties of harmonic measures, 2006
866 **William Craig,** Semigroups underlying first-order logic, 2006
865 **Nathanial P. Brown,** Invariant means and finite representation theory of $C*$-algebras, 2006
864 **John M. Lee,** Fredholm operators and Einstein metrics on conformally compact manifolds, 2006
863 **M. Lübke and A. Teleman,** The Universal Kobayashi-Hitchin correspondence on Hermitian manifolds, 2006
862 **Alberto Canonaco,** The Beilinson complex and canonical rings of irregular surfaces, 2006
861 **Leon A. Takhtajan and Lee-Peng Teo,** Weil-Petersson metric on the universal Teichmüller space, 2006
860 **Thomas M. Fiore,** Pseudo limits, biadjoints and pseudo algebras: Categorical foundations of conformal field theory, 2006
859 **N. Arcozzi, R. Rochberg, and E. Sawyer,** Carleson measures and interpolating sequences for Besov spaces on complex balls, 2006
858 **Enrico Valdinoci, Berardino Sciunzi, and Vasile Ovidiu Savin,** Flat level set regularity of p-Laplace phase transitions, 2006
857 **Donatella Danielli, Nocola Garofalo, and Duy-Minh Nhieu,** Non-doubling Ahlfors measures, perimeter measures, and the characterization of the trace spaces of Sobolev functions in Carnot-Carathéodory spaces, 2006
856 **Vladimir Bolotnikov and Harry Dym,** On boundary interpolation for matrix valued Schur functions, 2006
855 **Yevgenia Kashina, Yorck Sommerhäuser, and Yongchang Zhu,** On higher Frobenius-Schur indicators, 2006
854 **Noam Greenberg,** The role of true finiteness in the admissible recursively enumerable degrees, 2006
853 **Joachim Krieger,** Stability of spherically symmetric wave maps, 2006
852 **Viorel Barbu, Irena Lasiecka, and Roberto Triggiani,** Tangential boundary stabilization of Navier-Stokes equations, 2006
851 **Jie Wu,** On maps from loop suspensions to loop spaces and the shuffle relations on the Cohen groups, 2006

For a complete list of titles in this series, visit the
AMS Bookstore at **www.ams.org/bookstore/**.

WITHDRAWN